Ultralights

A Complete Introduction to the Revolutionary New Way to Fly

Ultralights

James E. Mrazek, Jr., UL2822
and
James E. Mrazek, Sr., UL2238

St. Martin's Press   New York

ULTRALIGHTS. Copyright © 1982 by James E. Mrazek, Jr., and James E. Mrazek, Sr. All rights reserved. Printed in the United States of America. No part of this book may be used or reproduced in any manner whatsoever without written permission except in the case of brief quotations embodied in critical articles or reviews. For information, address St. Martin's Press, 175 Fifth Avenue, New York, N.Y. 10010.

Design by Kingsley Parker.
Cover photo by Vince Streano, courtesy of Eipper Formance.

Library of Congress Cataloging in Publication Data

Mrazek, James E., 1944–
  Ultralights: a complete introduction to the revolutionary new way to fly.

  1. Ultralight aircraft.   I. Mrazek, James E.
II. Title.
TL671.M7   629.1′4   82-5788
ISBN 0-312-82852-7   AACR2
ISBN 0-312-82853-5 (pbk.)

First Edition

10 9 8 7 6 5 4 3 2 1

CONTENTS

	PREFACE	ix
1	EVOLUTION OF THE ULTRALIGHT	1

Hang Glider Regulation. The Powered Hang Glider. New Regulations. The Word "Ultralight" Evolves. Unique Phenomenon. Aviation's Best Bargain. Exceptional Growth. Good Hang Glider Sites Lacking. Appeal to Hang-Glider Flyers. Pilots, Flyers, and Operators. General Aviation's Concerns.

2 ULTRALIGHT CATEGORIES   11

Major Categories. Some Early Power Systems. Flexible-(Flex-) Wing Ultralights. The Trike. Rigid-Wing Ultralights. Hard-Wing Ultralights. Two Place Ultralights. Man- and Solar-Powered Ultralights. Ultralight Technical Data.

3 POWER SYSTEMS   43

Back-Pack Power. Recent Developments. Tractor Versus Pusher Power Systems. Advances in Australia. Development of Power Systems. Engine Facts. Two-Cycle Engines. Cooling. Design. Fuel. Starters. Reliability. Propeller (Prop). Direct Drive and Reduction Drive. Mufflers. Maintenance. Future Prospects. Engine Technical Data.

4 CONTROL SYSTEMS   62

Weight Shift Control. Weight Shift with Supplemental Aerodynamic Control(s). Three-Axis (Stick) Control. The Best System. Canards.

5 INSTRUMENTS AND GEAR   65

Airspeed Indicator. Altimeter. Variometer. Tachometer. Cylinder Head Temperature Gauge (CHT). Radio. Helmets. Fabric Care.

6 THEORY OF FLIGHT   72

The Ultralight in Flight. How a Wing Creates Lift. Angle of Attack and Stalls. The Three Axes. Turns. Glide Ratio. Wing Loading. Aspect Ratio.

7 LEARNING TO FLY THE ULTRALIGHT 81
Ultralight Flying Programs—Where to Find Out about Them. Manufacturers and Flying Programs. Training. Learning to Fly Ultralights. Ground Training. Preflight Inspection. Before Starting the Engine. Engine Start. Taxi. Takeoff. Cruise. Turns. Importance of Feel. Approach and Landing. After Landing. Postflight. Flight Maneuvers. Beware of Engine Failure. An Instructor is Best. Pilot Flight Log Book. Insurance.

8 PARACHUTES 99
Unique Circumstances. Parachutes in Use. Test the Parachute. Don't Wait! What Using a Parachute Can Mean. Further Development Needed. Evolving Systems. Care and Maintenance.

9 WIND AND WEATHER 106
Micrometeorology. The Weather Eye. Obtaining Weather Information. Highs and Lows. Winds. Minifacts about Micrometeorology. Wind Shear. Other Turbulence.

10 THE ULTRALIGHT AS A TOW AIRPLANE 117
The First Ultralight/Hang Glider Tow. Power and Towing. Ultralight Towing Advantages. Surplus Power Needed. Future Possibilities.

11 AMPHIBIOUS ULTRALIGHTS 121
Float Construction. Cautions. How They Fly. The Capsized Ultralight.

12 MANPOWERED AND SOLAR POWERED ULTRALIGHTS 129
The Hang Glider's Contribution. First Solar-Powered Flight. The "Forever Airplane."

13 HOW TO BUY AN ULTRALIGHT 134
"See It Before You Buy!" No Certification Required for Ultralights. PUMA. Picking a Dealer. Newly Introduced Models. Purchasing the Used Ultralight—Cautions. Building Your Own Ultralight. "Jury-Rig" Repairs. "Foot-Launchability." Avoid Pressure to Foot Launch.

14 MILITARY APPLICATIONS OF ULTRALIGHTS 143
Hit-and-Run Attacks. Active Combat Ultralights. Fighters.

15 CLUBS, ASSOCIATIONS, AND PUBLICATIONS 148
Clubs Form. Bay Area Ultralight Club. Getting Information. Powered Ultralight Manufacturer's Association (PUMA). Publications.

16 ROUNDUP OF ACTIVITIES AND ACHIEVEMENTS 153
Long-Distance Flights. Long-Duration Flights. Altitude Records. Ultralight Parks and Airports. Ultralight Activities Growing.

17 ULTRALIGHTS AROUND THE WORLD   159
    Canada. West Germany. United Kingdom. Japan.

18 THE FAA AND ULTRALIGHTS   169
    FAA Definition. Unforeseen Changes Occur. Irresponsible Acts. FAA Position. Pilot Certification. Aircraft Registration. Aircraft Certification. Rules. Possible Roles for Aircraft Owners and Pilots Associations. International Criteria. Ignorance of the Law? Regulation by States.

19 ABOUT THE FUTURE   179
    The Promise of Graphite. Costs and the Future of the Sport.

    ULTRALIGHT CLUB DIRECTORY   183

    GLOSSARY   186

    BIBLIOGRAPHY   193

PREFACE

The sport of ultralight flying today is in a state of dynamic change. It is young, exciting, challenging, and expanding rapidly with many new developments under way.

This book offers a broad survey. For the reader who desires to probe more deeply into a particular subject, additional readings have been suggested by the authors. Under no circumstances should one actively partake of this sport without first acquiring a thorough knowledge of it and properly building flying skills—constantly refining old skills and gradually adding new ones.

Many people in general aviation complain that some ultralight pilots inexcusably violate air regulations. The authors agree. All those who choose to participate in the sport should thoroughly understand pertinent federal and state regulations. Reading this book is no substitute for gaining a thorough grounding in flying skills and a knowledge of official regulations.

Ultralights

First flown on May 23, 1963, the Ryan "Flexwing" used an early design of the Rogallo flexible wing as a sail form. The *Flexwing* was the first powered hang glider. However, because it weighed 1,100 pounds, it would not meet the weight and foot launchability requirements set by the government for today's ultralights. Note the wing's high angle of attack and the deep "billow" of the sail. *Ryan official photo courtesy of Peter M. Bowers*

1 EVOLUTION OF THE ULTRALIGHT

In the late 1940s Dr. Frances Rogallo, a research scientist, and his wife Gertrude invented a flying wing that he referred to as a "simple limp wing form." A few years later, the National Aeronautic and Space Administration (NASA) experimented with a "para sail" based on Rogallo's wing, to glide space loads down to the earth. In the early 1960s the Army began experiments with a man-carrying powered Rogallo called the "Flex-Wing," which weighed 1,100 pounds. A conventional rudder controlled direction, but without aelerons, or a means for the pilot to shift his weight, wide sweeping turns had to be made to change direction. The successful flights of the Flex-Wing led the Army to contract for an aerodynamically cleaner design, the "Fleep."

Borrowing from features of these experimental craft, sports flight enthusiasts started towing the wing by boat with the pilot hanging in a harness below the wing. Soon daring pilots were launching the wing from hills and the almost forgotten art of powerless flight, called "hang gliding" or "sky sailing," was reborn. The limp or flexible wing took the name "Rogallo" hang glider after its inventor. Soon some "rigid" wing hang gliders appeared. Both the flexible Rogallo and rigid wing hang gliders played a crucial role in the flying revolution that was soon to occur with the appearance of the powered hang glider dubbed the "ultralight."

Hang Glider Regulation

Hang gliding grew in popularity and every weekend thousands were flying from hills and along ridges and sandy beaches across America. Unwittingly or otherwise, a few began to fly into regulated airspace endangering themselves and other aircraft. The Federal Aviation Administration (FAA) soon decided that some regulation of the sport was necessary. The rules (issued in 1974), contained a definition of a hang glider that was to have a serious impact on the development of the ultralight. It defined a hang glider as "an unpowered single place vehicle, whose *launch and landing depends on the legs of the occupant and whose ability to remain in flight is generated by the air currents only.*" However, the rules allowed hang glider pilots considerable freedom, leaving the regulation of the sport pretty much up to its members and their organizations. Under these conditions, the sport continued to enjoy a healthy growth.

The Powered Hang Glider

By late 1976, some pilots, dissatisfied with the limited glide range of their hang gliders, fitted their gliders with compact power systems using small engines from chain saws and outboard motors. Dubbed "glide extenders," pilots would start the engines while in flight to give a glide a little thrust, or to provide power between lifting thermals, or when other lift winds subsided. Others turned the engines on before launching, flying with power on throughout the flight. Soon some pilots equipped their craft with wheels, then ailerons and controls. With these additions, many powered hang gliders, especially the rigid wing models, looked very much like small, extremely (or "ultra" from the Latin word for extreme) lightweight airplanes.

New Regulations

The FAA noted the power trend with concern. Desirous that these new craft would not grow to a size and speed that would ultimately call for stringent regulation, the FAA, in 1978, broadened the scope of its defi-

Foot launching a *Pterodactyl Fledgling*. It is a Federal requirement that ultralights are foot launchable. *Photo courtesy of* Glider Rider

nition of the hang glider, describing it as either an "... unpowered or powered vehicle." It continued to describe the hang glider as a "... *foot launchable vehicle whose launch and landing depends solely on the legs of the operator,*" despite the fact that most pilots were no longer making running takeoffs nor landing on wheels. The FAA was aware of this but felt that including powered vehicles in the hang glider category with the retention of the foot launching requirement would tend to keep the craft lightweight, short flight, recreational, and free of onerous rules.

The Word "Ultralight" Evolves

In the 1970s the term "ultralight," which previously had been used only to describe certain lightweight aircraft, was applied to the hang glider. When powered models appeared, the name "ultralight," a word already acceptable to the sport, was quickly applied to the powered version and the many comparable lightweight craft that have since developed. Despite this turn of events, the FAA still refrains from calling powered hang gliders either airplanes or aircraft. It calls them "vehicles," although it appears that it may change this policy. One reason for using "vehicles" is that airplanes and other aircraft must be inspected and certified. Vehicles need not be.

The authors feel that the ultralight has met the essential criteria to fall within the accepted definition of "airplane," which according to Webster is "a fixed wing aircraft heavier than air that is driven . . . by a propeller . . . and supported by the dynamic reaction of air against its wings." In view of the uncertainty over what the FAA may ultimately decide to call the ultralight, the authors will refer to ultralights in this book as "ultralights," and also as "aircraft," or simply "craft."

Unique Phenomenon

The ultralight phenomenon is unique: no single element of general aviation can lay claim to its birth and growth. No corporate giant in aviation sponsors it. Instead, the ultralight has bewildered these giants, who are wondering where this whippersnapper came from so suddenly.

In America, wealthy as well as shoestring pilots have been buying ultralights. One reason for their choice is the "hassle factor." Recreational or private flying in experimental or other small airplanes is subject to more rules and regulations than any other leisure activity in America. Ultralight flying on the other hand has remained relatively regulation-free.

Aviation's Best Bargain

The greatest barrier to the popularization of private flying before the advent of the ultralight was cost. In recent years, it cost a minimum of two thousand dollars at thirty-five dollars an hour, plus fifteen dollars an hour to rent a plane to earn a private pilot license. Cost undoubtedly discouraged more people from flying than the imagined difficulty of learning to fly, the hours required to learn regulations, the risks, or any other factors.

"For all that money," one ultralight enthusiast says in *Glider Rider*, "the reward is a piece of paper that says 'private pilot, single engine.'" He cites the expenses of buying a plane, paying yearly taxes, tie-down or hangar fees, and FAA-mandated annual inspections as added costs.

Randall Brink, the editor of *Ultralight Flyer*, believes that the high cost of flying light airplanes in the past made it difficult to truly enjoy recreational flying. However, he states: "We suddenly have a lot more going for us ... because flying enthusiasts and innovators got busy and did what the major aircraft manufacturers have continually tried and failed to do: Build simple, fun flying machines; ones that we can afford to buy and fly: Ultralights. These are airplanes that are so simple, and there is no need for the licenses and restrictions that have developed cheek by jowl with the advancements of 'conventional' airplanes. Ultralights are easily maintained," he says, without the "... aid of expensive certificated aircraft mechanics."

Exceptional Growth

Current production statistics indicate that many features of ultralights are proving attractive to potential and active flyers. According to Michael A. Markowski, author and consultant on ultralights, in 1981 ultralight manufacturers built more than ten thousand craft, to exceed the total number produced by all manufacturers of airplanes for general aviation. He estimates that ultralight production will exceed twenty thousand units in 1982! By the authors' estimates this should make it a $120,000,000 industry in 1982. These impressive results are the achievement of a group of young entrepreneurs who, on average, have been in business only two and a half years. If Eipper Formance, manufacturer of the *Quicksilvers,* meets its planned production objective of thirty-six hundred units in 1982, it will be the largest producer of aircraft in the world! Impressive performances indeed, considering that the first ultralight was manufactured in 1975.

Good Hang Glider Sites Lacking

There are several factors that favor the ultralight sport over the sport of hang gliding. While many hang gliding sites exist in the United States and other nations, many current and potential sites are not as good as they might first appear. John Moody, a hang glider and ultralight pilot, quipped in an article appearing in the February 1976 issue of *Sports Aviation,* "A hill that looks usable a mile away, invariably when you get up to it, has six fences across its face, has a resident herd of bulls who have not seen a cow in six months, and a hundred and fifty acres of briars at its base." Such experiences are common in the sport.

Many hang-glider flyers living close to good hang-glider launching sites would like the option of flying a recreational powered aircraft, particularly at times when the weather is not good for hang gliding. Of course, most people around the world do not live close to mountain- or ocean-hang-gliding-and-soaring sites. Ultralights offer them what hang gliders offer others, and perhaps something more—controlled powered flight. Many have dreamed all their lives of someday flying an airplane. Ultralight flying can be a way for them to fulfill their dreams.

In addition, many people who want to fly have considered the hang glider an inexpensive and useful tool for satisfying this yen. But people like this shy away from learning to hang glide for one major reason: they see that, sooner or later, in order to enjoy this method of flight, they will have to launch from a 3,000- to 5,000-foot cliff—a frightening possibility. On the other hand, flying with power up to that same height or higher does not faze them.

Grounded hang gliders. The dilemma of a hang glider flyer—waiting for a lift wind.

Appeal to Hang-Glider Flyers

Despite the many appeals of hang gliding, most flyers feel no urge to try such engineless flight. The inconvenience of the sport is one reason for their disinterest. When winds are inadequate, or there is no thermal lift to carry hang gliders into the air, gliding or soaring is not possible, and pilots are grounded. Ultralights answer this frequent dilemma by giving the hang-glider pilot hours at a time in the air and the ability to fly in any direction, among other advantages. Ultralights also give their pilots an extra bonus when entering lift. Pilots can turn off their engines and glide or soar—just as if they were flying in an engineless hang glider, but without the risks.

Pilots, Flyers, and Operators

Some ultralight flyers have private or commercial pilots' licenses. The majority do not, and there is no present requirement that they earn a license to fly an ultralight although the FAA may soon change this.

The FAA refers to pilots of these aircraft as "operators," and the aircraft as "vehicles." In three respects the FAA distinguishes the sport: Ultralight pilots are "operators," "ultralights" are "vehicles," and "ultralights" are foot-launchable aircraft. All terms are carryovers from the sport of hang gliding.

Because of the FAA viewpoint, and the lack of licensing of many ultralight pilots, a current widely held opinion is that those who fly ultralights and are not licensed pilots should not be called "pilots." The authors feel otherwise. Ultralights fall within Webster's definition of an airplane as an "... engine-driven, fixed-wing aircraft heavier than air that is supported in flight by the dynamic reaction of the air against its wings." Moreover, Webster defines "pilot" as a person "... qualified to fly an airplane." Thus, any person who flys an ultralight must be considered as much of a pilot as a person flying a Boeing 747, albeit, much of the training and flying skills are considerably different.

Since flying an ultralight requires dexterity and skill, in some respects greater than that needed to fly many other aircraft, the authors will usually refer to flyers of ultralights not as "operators" but as "pilots," recognizing that they have the skills to qualify as such.

General Aviation's Concerns

With the addition of engines, landing wheels and other gear to 35- to 50-pound hang gliders, the weight of these hang gliders converted to power has tripled. This much heavier weight began to erode the ability of the pilot to foot launch.

The conventional method for taking off in an ultralight is now on wheels. Increased weight, coupled with the disappearance of the foot-launch characteristic, tends to place the ultralight in the class of a light airplane.

Lately, ultralights have begun to be used to tow advertising banners along ocean beaches, to dust crops and make other commercial flights, thereby taking them out of the sport category.

Jet Wing All Terrain Vehicle (ATV). The 37-horsepower "trike" is bolted to a *Demon* flexible-wing hang glider. *Photo by David A. Gustafson*

2 ULTRALIGHT CATEGORIES

Ultralights (powered hang gliders) have common characteristics that set them apart as a class of aircraft. The major ones are

- their designs are based primarily on the early flexible and rigid wing hang gliders
- low cruising speeds of about 45 miles per hour
- they weigh about 254 pounds (empty)
- their wing loading is about 2 pounds per square foot of wing area
- they have low powered (20 to 30 horsepower) two cycle engines
- they do not have to be inspected or certified by the FAA
- their pilots need not be licensed
- they are limited to speeds of less than 55 knots
- they carry no more than 5 gallons (30 pounds) of fuel

Ultralights are grouped (separated) in two ways. One is the manner by which flight is controlled, the other by their design. There are essentially two methods of control used. One is by weight shift, the other by conventional aerodynamic controls. Those presented in this chapter will not be separated according to their control systems, but rather by design category.

Major Categories

Ultralights around the world can generally be separated into categories. Some may fall into more than one category. They are:

- flexible wing
- rigid wing
- hard wing

While it is expected that the FAA will modify the parameters that govern the designs of ultralights, it is felt that these categories will still pertain to the crop of new designs that certainly will appear.

Photographs of several lightweight aircraft that do not have ultralight characteristics described above will be found on these pages. They are included because they show interesting trends or details.

Ultralights are sold primarily in kits by manufacturers and dealers. Buying a kit saves the purchaser manufacturing and shipping costs, but involves the commitment of the time required to build the ultralight. Kits cost from $3,500 to $5,500 depending on the ultralight model. In certain models a purchaser can buy a ready-to-fly ultralight from dealers at a markup of $500 to $800 over the kit price. The dealer charges this premium to assemble the ultralight. Dealers usually test fly these machines thoroughly before they turn them over to their new owners. More about kits is contained in Chapter 13.

The ultralights discussed in this chapter have wing loadings averaging almost 3 pounds. Some airplanes with wing loadings in excess of 3 pounds are also described as ultralights. A good example is the waist-high, home built French *Cri Cri* (from cricket). It is lightweight (only 160 pounds) because of its narrow and very short wing (giving it a small wing area) but its wing loading is 11 pounds, or about four times that of the conventional ultralight. There is additional discussion about the *Cri Cri* and several comparable airplanes in Chapter 18.

Some "home-built," or "experimental" airplanes are called ultralights because they are very lightweight.

In spite of its light weight (160 pounds), the French-designed *Cricket* (Cri Cri in French) is not considered an ultralight. One reason is that it is not foot launchable. Another is its 11-pound wing loading. It must be certified and its pilot must be licensed. *Photo courtesy of Zenair, Ltd.*

They do not require certification in the usual sense but their pilots must be licensed.

The FAA applies the term "ultralight" to only a narrow class of aircraft. This intention will most likely continue when the FAA publishes its new rules and guidelines.

Some Early Power Systems

Some of the earliest efforts to build powered hang gliders, the forerunners of today's ultralights, took place in 1976. One of the more interesting endeavors was the use of a backpack power system (described in more detail in Chapter 3). Several hundred of these units were sold to be used primarily with flex-wing hang gliders. Pilots also converted chainsaw motors, outboard motors, and other small combustion engines into power systems for hang gliders. Many of these later engines flew remarkably well, and some can still be seen in the skies. These models proved better than the backpack models, which soon disappeared, although occasionally an adventurous pilot still uses one.

Stimulated by the interest in powered hang gliders, several manufacturers began to produce power systems especially for flex-wing hang gliders. The well-

Soarmaster PP106 power system installed along the keel of a *Cirrus* flexible-wing hang glider. This simple, easy-to-install system using a two-stroke, 10-horsepower, Chrysler engine, can occasionally be seen flying in the United States and other nations. *Photo courtesy of Soarmaster, Inc.*

known, 30 pound PP-106-power system developed by Soarmaster is one. This system attaches to the keel of the glider. The drive shaft extends to the rear, and the propeller protrudes some two feet beyond the trailing edge of the wing (see photo, above). In spite of its attractive low cost of approximately $1,000, this unit did not catch on in the sport, and is seldom seen in use in America because of a rash of accidents, although foreign exports and the development of similar systems in other countries indicates that the economy and easy installation, plus other factors are recognized as worthwhile to the sport's needs and interests.

Flexible- (Flex-) Wing Ultralights

Originally, a dominant feature of flex-wing aircraft was the absence of a rigidly built wing cross-section (airfoil) to cause the Dacron sail to take the shape of an efficient lift producer. Instead of using an airfoil, early flex wings relied on the wind to inflate their sails into an airfoil shape. Today, airfoil-shaped battens are used in flex wings to stiffen the sail and have eliminated this distinction. A major feature remains, however. While the flex wings all have a rigid aluminum tube along the front, or leading edge, the rear, or trailing edge, is flexi-

The *Eagle* is one of the few flexible-wing ultralights. Note that it has a small leading wing (canard) that has an elevator attached to its trailing edge. Dropping below the wing tips are rudders. *Photo by the authors.*

ble, sagging when the glider is on the ground and moving flexibly in response to wind forces when the glider is in flight. This flexibility typifies most hang gliders with the notable exception of the *Fledgling*, the *Quicksilver,* and the *Mitchell Wing.* Another characteristic of the flex wing is that it can be quickly taken apart, packed into a compact package, and transported on the top of a car. Many fit into a 12 foot bag that is only 10 inches in diameter.

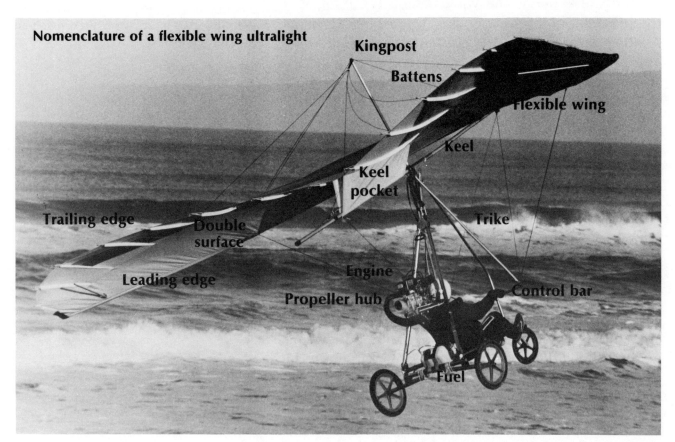

Nomenclature of a flexible wing ultralight

The flex wing's battens are usually preformed, air-foil-shaped, plastic stays (see photo, opposite) that slip into wing sleeves. These act as stiffeners for the wing, but are not rigid in the way that aluminum tubing, wood, or plastic-rib airfoils are.

The most popular flex-wing ultralight especially designed as an ultralight (rather than adapted from a hang glider) is the weight shift controlled *Eagle* (see photo, page 17). It has a 36-foot-long, 188-square-foot wing. It weighs 163 pounds and can carry a pilot weighing up to 300 pounds. The standard engine is the single cylinder, 20-horsepower Zenoah.

The *Eagle* has a wing, leading the main wing, called a "canard." It improves the stability of the craft in flight. Rudders drop below the wing tips. Elevators are on the canard. Cables connect the elevator to the seat. As the pilot moves the seat backward, the elevator tips up and the craft rises. Shifting forward causes the elevators to lower, and tips the nose downward.

The rudders are connected to a steering bar. This controls yaw/roll and changes the direction of flight. The engine is below the wing behind the pilot, with the propeller behind the engine. The *Eagle* can be assembled or dismantled in about one-and-a-half hours. The airframe folds into an 11-foot-long package that is 10 inches in diameter, making it easy to carry on a car roof. Of all ultralight models, it is the most compact when disassembled, a feature that adds to this ultralight's popularity.

Pilot holding three preformed wing battens. Inserted in the wing, they give it an air-foil shape.

The Trike

A system imported from England that is gaining in popularity in the United States and elsewhere, is the "trike," a bolt-on unit that converts a hang glider into an ultralight in a matter of minutes. It is a self contained, seven foot high, pyramidal looking "pod," or "cage," that houses the pilot, power system, fuel tank, and landing wheels. The pilot sits in the pod and maneuvers the glider by shifting the control frame to his front (see photo, page 10).

A trike powered hang glider is flown strictly by weight shift, therefore, a good deal of hang gliding ex-

Eagle flying above an automobile carrying another *Eagle* that has been disassembled, packaged, and secured to the top of a car for transport. The *Eagle* can be unpacked and readied for flight in 15 minutes by two men, and readied for transport in less time. *Photo courtesy of American Aerolights*

perience prior to flying is necessary, and even an experienced hang glider pilot finds his skills tested by the special flying characteristics of a trike. A benefit of the trike ultralight is that hang gliders manufactured in America, are most likely certified by the United States Hang Glider Association's Manufacturer's Association, and have met standards set by the latter. There is no similar certification as yet set up for ultralights in general.

For owners of hang gliders, a trike provides an inexpensive way to become an ultralight owner and enjoy the dual advantages of pure hang gliding and powered ultralight flying at low cost. The average trike costs $1,800; the average, ready-to-fly ultralight costs about $5,500. Thus, the pilot who already owns a hang glider can turn it into an ultralight at about a third the cost of a new ultralight.

Those who do not own hang gliders can purchase a new one for approximately $1,700, or a used one for as little as $500 to $700, and then buy a $1,800 trike; thus, the total cost of a trike ultralight would range from $2,200 to $3,500, considerably below the cost of either the rigid wing or hard wing ultralight, and would provide a very economical way of enjoying the sport.

When a hang-glider pilot finds the winds are not favorable for hang gliding, or has a yen for flying an airplane for a few hours, he transports the folded hang glider and the trike to a takeoff strip, unfolds the glider, bolts on the trike to the glider's keel and has an ultralight airplane ready to fly.

The pilot sits within the trike frame below and behind the control bar. The pilot shifts the trike forward or backward, or to either side with reference to the control bar, to alter the thrust line and control the craft's flight. Although the trike is used by many pilots in England, where it originated, so far it has had only a limited acceptance in the United States. Pilots are finding that despite their proven skill in flying a hang glider and one or several ultra models, flying in a trike involves them in control operations quite different from any others. New skills unlike anything else in aviation must be learned and perfected.

It will be some time before there is enough information to permit a well-rounded approach to a trike's operation. Meanwhile, those who would consider flying one are advised to do a good deal of research and heed George Christian Marechal's article, "From Rag Wings to Riches," November 1981 issue of *Glider Rider,* which states, "The trike-powered hang glider was okay for what it was. But piloting one for 35 hours on the Land's End race taught me what, in retrospect, was one of my most valuable lessons: You've got to be a hang-glider pilot to be able to learn to fly a trike with reasonable safety." The authors contend that it takes not just a "hang glider pilot," but a *highly experienced* hang glider pilot to fly one *safely.*

Monowing

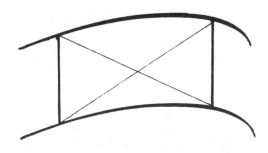

Biwing (biplane)

Rigid-Wing Ultralights

The main feature of a rigid-wing ultralight is that the Dacron wing has a rigid leading edge, usually an aluminum tube, and a fixed (rigid) trailing edge. The Dacron may, or may not be stiffened by doping it. The wing may have a single surface, a partial double surface, or a double surface. The upper surface may be stiffened with ribs or battens to give it an airfoil shape. All rigid wings produced are single wing monoplanes.

The *Mirage,* a three-axis, conventional-control, rigid-wing ultralight. Note droop tips on the ends of the wing to reduce drag caused by wing vortex turbulence. *Photo courtesy of* Glider Rider

The *Snoop* is a rigid-wing ultralight that has three-axis controls and a wide flying cage to accommodate extra payloads. *Photo Courtesy Eastern Ultralights*

The *Weedhopper* is a rigid-wing, three-axis control ultralight powered by the Chotia engine, manufactured by the builder of the *Weedhopper*. *Photo by the authors*

Nomenclature of a rigid-wing ultralight

Photo by Steven McCarroll; Courtesy of Eipper Formance

Semi-Rigid Wing Ultralights

The distinguishing features of this category are the wing's rigid leading and trailing edges, and its single surface (see illustration). This surface may have stiffeners (battens) inserted.

Prominent among single surface rigid wings are the powered *Quicksilvers*. Their design is based on the airframe of the popular and successful *Quicksilver* hang glider first flown in the early 1970s and well known for its stability and reliability. Some enterprising owners in the late 1970s attached power systems to their "*Quicks*." They flew surprisingly well. Eipper Formance, the manufacturer of the *Quicksilver* hang glider, soon took the cue and designed a powered version in 1979 (see photo, below). Because the *Quicksilver* hang glider and its powered successor have been reliable and stable, there have been many imitations (known in the sport as "clones") of the basic airframe design, as is apparent in the photographs of many of the ultralight models in this book.

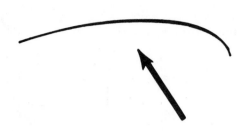

No bottom to air foil

Single surface

Quicksilver MX (multi-axis control) above; *Quicksilver* weight-shift control model below. The narrow white flaps across the colored stripes are spoilerons. They are activated by the foot pedals and can be used individually as ailerons for roll control, or together as spoilers for glide rate control, particularly in landing descent. *Photo courtesy of Eipper Formance*

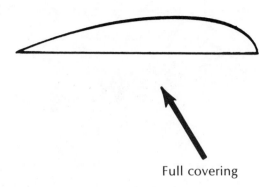

Full covering

Double surface

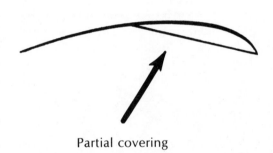

Partial covering

Partial double surface

The swing-seat, weight-shift controlled *Quicksilver* is a single-surface fixed-wing, single-engine monoplane with a conventional king post and cable bracing. It has a 32-foot span, 160-square-foot wing, weighs 156 pounds, and can carry a pilot weighing as much as 230 pounds. The engine is suspended under the wing center with the drive shaft extending to the propeller behind the wing. The standard motor is the single cylinder Cuyuna 215. Elevators attached to the stabilizer are moved by the pilot shifting the seat. Yaw and roll are obtained by weight shift augmented by the rudder, which is connected to the pilot's harness. The *Quicksilver* is quickly assembled and disassembled into a 1 X 5 X 16-foot package, including the engine and propeller, and can be carried on a car roof.

Eipper Formance now builds a three-axis-controlled model, the *Quicksilver MX*, designed to satisfy the demands of many pilots who want the economy and ease of flying an ultralight but who prefer standard aircraft controls instead of a weight-shift system. It features spoilerons above the wing, which are operated by foot pedals. Power is supplied by a 30-horsepower Cuyuna 430D that turns a 52-inch hardwood propeller. The *MX* weighs 220 pounds.

Other single-surface-wing ultralights may have some similarities to, such as single-surface wings, but many differ in other prominent design features. *Teratron*, for example, has a low initial cost, very low noise, low power, and slow speed—it is not a flashy machine, but these qualities appeal to many novice ultralight flyers.

Double-surface, semirigid-wing ultralights have wings with a top and bottom surface. The top surface in many has airfoil-shaped ribs that are, in some cases, removable. In some, the double surface extends only partway under the wing from the leading edge (see illustrations on left). Prominent among the double-surface, semirigid-wing ultralights are the *Pterodactyl Fledge, Hummingbird, Hummer* and *Mirage,* shown in the accompanying photos on pages 22 and 23.

The *Pterodactyl* is the well known *Manta Fledgling* hang glider fitted with a power system. The wing has a 32-foot span and a 162-square foot area. It weighs 170 pounds and can carry a pilot weighing as much as 250 pounds. It uses a Cuyuna 430D engine. It flies by supplemental weight shift and has a control stick, or twist grips, for yaw and roll obtained by means of drag-tip rudders above the wings. The stick gives "up" pitch only—"down" is by weight shift and trim.

The engine is at the rear, below the wing, and is direct drive to the 36 inch diameter, 16 inch pitch propeller ("pitch," theoretically, is the distance a propeller will advance in one revolution). Packed for transport, the craft can be carried atop a car.

The *Pterodactyl Fledgling* (Pfledge), an early and popular ultralight that uses the *Manta Fledgling* hang glider for its wing. *Photo above by Dan Johnson, courtesy of* Whole Air *photo left by the authors*

The *Pterodactyl Traveler* (Ptraveler) has a pivoting canard that serves as an elevator. The winglets operate as rudders and spoilers. The *Pterodactyl Ascender* (not shown) is the same as the *Traveler*, but uses a reduction drive power system.

Two views of the *Hummingbird*. The partially enclosed fuselage is for inclement weather flying. Note plastic "see-through" screen protecting pilot. V-tail with "ruddervators" (combination of rudder and elevator) plus ailerons provide for three-axis control. The Gemini/Partner K1200 Swedish chain saw engines turn out 8 horsepower each, and together "sip" *less* than a gallon of fuel per hour of flight. Fuel cost is thus about $1 for an hour of flying. *Hummingbird* has recently been equipped with twin-cylinder Limbachs on its enclosed fuselage prospector, upping its performance considerably. *Photos by the authors*

The rigid-wing *Hummer* has conventional three-axis controls and a "V" tail. It is one of the early designs and is popular and reliable. *Photo by the authors*

Hard-Wing Ultralights

"Hard"-wing ultralights have characteristics and components similar to those found in conventional airplanes. Depending on the ultralight, these may include doped fabric, or hard fiberglass, plywood, or aluminum wing surfaces that are in some cases Mylar® glued and heat shrunk to the base structure of the wing. The airfoil cross section of the wing is maintained by rigid wood or styrofoam ribs. The wings may be supported by struts or wire supports, or they may be unsupported and fully cantilevered (unsupported) as are the wings of the *Mitchell B-10* and the *Goldwing*.

Wing of the hard-wing *Mitchell B10*

Note the rigidly constructed airfoils and "D" cross-section numbers of the main spar, characteristic of hard-wing ultralights.

The *Mitchell B-10* is in a class by itself because of its speed and advanced design. It can cruise at 55 miles per hour! It is a wood-and-fabric flying wing with fully cantilevered, tapered, rigid wings, and a fully enclosed cabin. The wing has a 34-foot span and a 136-square-foot area. The craft weighs 170 pounds and can carry a pilot weighing as much as 280 pounds.

The *Mitchell* has three-axis aerodynamic controls using a stick and rudder pedal. The power systems use either the Zenoah, Honda, or Cuyuna engines and are located behind the cockpit. If the pilot is looking for speed and ability to contend with rough air, then the

The *Mitchell U-2* has a hard, fully cantilevered (unsupported) wing with wing tip rudders. It has conventional, three-axis controls, and cruises at more than 65 miles per hour. The cockpit is enclosed. An interesting feature of the *Mitchell* ultralights is their low profile. *Photo by the authors*

Based on the well known *Mitchell* flying-wing hang glider, the fully-cantilevered *Mitchell B 10* has three-axis controls. It has the same wing as the *Mitchell U-2* and similarly has wing tip rudders. It cruises at more than 50 miles per hour. *Photo by the authors*

Twin-boom, twin-fin, hard-wing *Mitchell P38*. The wing can be folded upward on both sides at junction of the wing strut for the purpose of transporting or storing the craft. As with the other *Mitchell* models, the engine is in the rear, making it a "pusher" power system. *Photo by the authors*

Mitchell B-10 would be the kind of ultralight he or she might like. However, this model takes a considerable number of hours to build and requires more storage space than many other models do.

Built in Canada, the *Lazair* is a twin-engine, rigid-wing monoplane with an inverted V-tail with tapered strutted wings (see photo, page 26). The wing span is 36.6 feet. The craft weighs 145 pounds. The wing has upswept tips and a 142-square-foot area. The *Lazair* has three-axis controls. The control stick is overhead. Yaw and roll are coordinated through a "mixer" situated between the wings. It has two Pioneer P60 (modified chainsaw) 5.5-horsepower engines. In addition, the craft has low drag and a high aspect ratio wing producing a thirteen to one glide, claimed to be higher than that of any other ultralight. Because of this excellent glide ratio, many *Lazair* owners routinely shut down their engines when at sufficient altitude, and

Manufactured in Canada, the translucent, unusually designed, efficient *Lazair*. Note the unconventional inverted "V" tail. The wing covering is Mylar, the engines, Pioneer chain saw. *Photo courtesy of* Glider Rider

The *Easy Riser* ultralight is the earliest and best known of the biplane ultralights, and the powered version of the famous *Easy Riser* hang glider. This model has a "caster" landing gear in which all wheels can be turned towards the desired direction of travel, permitting the wing to stay facing the wind, thus making it excellent for cross wind landings. It also makes taxiing in a wind easier. This model uses the Moody/Stewart designed "Maximizer," a two-engine, two-propeller, counter-rotating-propeller system shown below.
a) *Photo, facing page, by the authors*
b) *Photo above by Dan Johnson, courtesy of* Whole Air

soar. The latest model *Lazair* has 9.5 horsepower Rotex engines, downswept wing tips and is 38 pounds heavier.

An excellent example of a biplane ultralight is the *Easy Riser*, a tailless, rigid-wing plane flown by supplemental weight shift. It has a 30-foot wing with a 170-square-foot area. It weighs 120 pounds not including the engine, and can carry a pilot weighing 230 pounds. For the *Easy Riser*, CGS Aviation provides a power pack consisting of a 22-horsepower engine and a single Ritz 54-inch X 30-inch wood propeller. Stewart Products furnishes a twin-engine power system with two counter-rotating concentric propellers.

Other interesting ultralights are shown in the accompanying photo section.

Co-author James E. Mrazek, Sr. at the controls of the rigid-wing *Tomcat*, an unconventional design even for an ultralight. It was awarded first prize by the Experimental Aircraft Association at Oshkosh, Wisconsin, in 1981, for the most outstanding new design. Angled vanes under the wing called "dehedral stabilizers" give lift, stability, and automatic role. The "all moving" canard tilts up, down, and sideways, and serves as both a rudder and an elevator. *Photo by Joel Strait*

Flight Design's rigid-wing *440 ST* first introduced at the Lakeland, Florida, Sun 'n Fun in March 1982, has twin booms, and a ballistically-deployable parachute. The chute is enclosed in a module that can be seen on the top of the wing. The wheels, parachute cannister and other members are faired (streamlined) to lessen drag. *Photo by the authors*

The Robertson *BI-RD* (frequently referred to as the *Bird*) has three-axis controls. It was first introduced at the Sun 'n Fun in Lakeland, Florida, in March 1982. Its distinguishing feature is the full-length, slotted ailerons that drop below the trailing edge of the wing. As with a number of other ultralights, the *Bird* is a tail dragger; that is, its tail rests on the ground when taxiing, or when parked. *Photo courtesy of Robertson Aircraft Corporation*

Startlingly different in concept, the tailless *Kasperwing* is one of the few ultralights that still uses weight shift for pitch control. Its ingenious reflex wing system, said to generate vortex lift, is combined with pivoting "winglets" that can be deployed individually as rudders for yaw control or collectively as spoilers. *Photo courtesy of* Glider Rider

Dale Kjesllsen flies the *Teratorn* (Greek for wonder bird), which he manufactures. Control of the above model is by weight shift. The new *Teratorn TA* has conventional three-axis controls and weighs 180 pounds, 25 pounds more than the weight-shift model. Both models are powered by the Rotax engine of Austrian manufacture. *Photo by Dan Johnson, courtesy of* Whole Air

Ultralight *Flights' Phantom* (below) introduced at the Sun 'n Fun in Florida in March 1982, has an enclosed fuselage, wheel "pants", a "crucifix" empennage, and three-axis controls. *Photo by Joel Strait*

Introduced at the Sun 'n Fun in Lakeland, Florida, in March 1982, the *Swallow* (above) has conventional controls and control surfaces. Note that the crucifix tail is supported by twin booms. *Photo courtesy of Swallow Airplane Company*

The rigid-wing *Sunburst* has an inverted "V" tail. Many claim the "V" tail is the most efficient empennage. *Photo by the authors*

The rigid wing *Vector 600*, a Klaus Hill design, originally called the *Humbug*. It has a "V" tail, ruddervators, and spoilerons, and has three-axis controls. It is a pusher-power arrangement (engine in front, propeller behind the wing), a feature found in several other ultralights. *Photo by the authors*

Rally 2B is a 3-axis control rigid wing craft. The center-mounted overhead control stick is an unusual feature. *Photo courtesy of ROTEC by Scott Reglan*

Two different models of the hard-wing *Invader*. Note the styrofoam air foils showing through the Mylar wing covering. The *Invader* was first introduced at the Sun 'n Fun in Lakeland, Florida, in March 1982. It has a steerable nose wheel and a full flying "V" tail, each side of which pivots in the control operation. It is available in plans only, and takes 400 hours to construct, according to its designer. Kits are being developed. *Photos by the authors*

The *Kolb Flyer* is an early ultralight design. It is a cleverly designed hard wing that uses welded construction to produce a lightweight (155-pound) aircraft. It is powered by Solo engines and has conventional, three-axis controls. *Photo by the authors*

The *DSQ Nomad* is a conventional three-axis controlled, hard-wing ultralight. Its unusual features consist of a single landing wheel, a "crucifix" tail and a single overhead boom. Note the support structure for the boom. *Photo courtesy of Glider Rider*

Oblique view of the *CGS Hawk* showing the interior. The *Hawk* was first introduced at the Sun 'n Fun in Lakeland, Florida in March 1982. The cockpit is canvas enclosed, the craft has three-axis controls and full flaps. It is referred to as the "Piper J3 Cub" of the ultralight industry, and is a departure in ultralight design comparable in many respects to a conventional airplane, using proven aerodynamic principles. It has a control stick in the middle of the forward part of the cockpit and pedals to operate the controls. It cruises from 40 to 60 miles per hour, and costs about $3 an hour to fly. It is being marketed in kits. *Upper photo by the authors, lower photo by Joel Strait*

American Aerolights' experimental hard-wing *Falcon*, developed from their *Eagle*. It has an enclosed cockpit, a nose canard, and a pusher-power system. *Photo by the authors*

Another view of the *Falcon. Photo by Joel Strait*

The *Papillon*, a "registered" aircraft ultralight. Note the "V" tail and the poly-dihedral wing, multiple since it bends upward at a dihedral in the center, and also at about midpoint between the center and each wing tip. It is three-axis controlled. *Photo by Dan Johnson, courtesy of* Whole Air

The *Rally 2B* is a pusher-propeller, twin-boom, high-wing externally-braced monoplane. It has conventional three-axis controls. Note the overhead control stick. *Photo by Scott M. Riglan*

Preformed Ultralights

The most prominent ultralight in this class at this time is the *Goldwing* (see photo). A space-age airplane in many respects, the *Goldwing* is a canard, single-engine craft with three-axis controls including stick-control spoilers, ailerons and elevators, and pedal control rudders. The double-surface wing is of preformed fiberglass and foam. It is 30 feet long and has an area of 128 square feet. The craft weighs 185 pounds and can carry a pilot weighing as much as 240 pounds. Its twin-cylinder Cuyuna 430 engine sits behind the cockpit and has a direct drive shaft to the 36 X 16-inch wood propeller.

Another lightweight aircraft that compares with the *Goldwing* is the *Striplin* (many do not consider the latter an ultralight). Other craft that border on being light fixed-wing airplanes are the *Cri Cri*, discussed earlier, and the *Quickie*.

Two Place Ultralights

Some feel that a student pilot should have some training in a two-place ultralight trainer before flying alone. Manufacturers have introduced two-seat models. None, so far, have been certified by the FAA. If they are certified, it is expected that they will be flown

A canard design, with wing tip rudders; the *Goldwing* is a fiberglass hard-wing ultralight. It is foot launchable and has conventional, three-axis controls. Its cruising speed is 75 miles per hour, or about twice that of the average flexible and rigid-wing ultralights. *Photo courtesy of Goldwing, Ltd.*

by a licensed pilot and not flown for commercial purposes. Thus, a dealer presumably might take a student for a "familiarization flight," but could not charge the student for the ride.

Man- and Solar-Powered Ultralights

The man-powered and solar-powered ultralights are unique even in a sport filled with unique craft and accomplishments. They are discussed at length in Chapter 12.

Wizard's prototype two-seat, dual-control ultralight introduced at the Sun 'n Fun in Lakeland, Florida, in March 1982. Under present government rules, the pilot of a two seat ultralight must have a pilot's license and the craft must have a certificate. *Photo by the authors*

Two-place, twin-engine, hard-wing *Sky Ranger* has a fiberglass composite construction. Its wingspan is 35 feet and wing area 158 square feet. It weighs 216 pounds and although possessing many of the characteristics of a true ultralight, cannot be foot launched, and thus falls between an ultralight and a lightweight airplane. It has conventional three-axis controls, is semi-aerobatic, and cruises at 85 miles an hour. The manufacturer recommends that the pilot be licensed. *Photo by Dan Johnson, courtesy of Whole Air*

39

TECHNICAL DATA

ULTRALIGHT MODEL   WING

	Span (feet)	Area (square feet)	Aspect ratio	Wing loading (pounds per square foot)
Flexible Wing				
Eagle	36	188	6.7	1.8
Jetwing ATV	33	176	6.2	N/A
Rigid Wing				
Cloudbuster	36	149	8.7	2.8
Condor	33	160	N/A	2.1
Hawk	29	145	5.6	
Hi-Nuski Huski	34	169	6.8	2.1
Hummer	34	128	7.8	2.7
Hummingbird	34	158	7.5	2.1
Mac 300	32	167	7.1	3.0
Mirage	34	140	8.5	2.5
Pterodactyl Fledgling	33	162	6.6	2.1
Quicksilver	32	160	6.4	2.0
Quicksilver MX	32	160	6.4	2.3
Rally 2B	32	155	6.4	1.9
Robertson B1-RD	30	162		2.7
Sunburst	36	168	7.7	2.1
Swallow	34	136	N/A	2.6
Teratron	32	160	6.4	2.0
Tomcat	30	175	N/A	2.2
Weedhopper	28	168	4.7	1.9
Wizard	32	160	6.4	2.0
Vector 600	34	160	8.5	2.1
Hard Wing				
Easy Riser	30	170	8.8	1.7
Goldwing	30	128	7.5	3.2
Invader	31	140	6.9	2.4
Kolb Flyer	30	160	5.5	2.0
Lazair	36	142	9.3	2.2
Mitchell B-10	34	136	8.5	2.3
Nomad DS-2GA	36	147	8.8	2.2

* Note: "Empty" weight as taken from the manufacturer's specifications and given here generally includes accessories, fuel, and optional equipment. However, "Wing loading" above is calculated as "Empty" weight minus fuel weight,

	WEIGHTS		ENGINE	CONTROLS		
Empty*	Pilot maximum weight	Maximum gross weight		Weight shift	Multi-axis	Combination

Flexible Wing

163	300	463	Zenoah	X		X
N/A	235	N/A	Kawasaki 44A	X		

Rigid Wing

248	275	532	Cuyuna 215R		X	
164	N/A	N/A	Kawasaki		X	
225	N/A	N/A	CGS Powerhawk		X	
190	250	440	Cuyuna 430D	X		X
175	240	420	Zenoah		X	
163	250	413	Gemini/Partner		X	
215	315	525				
185	280	465	Kawasaki		X	
170	250	420	Cuyuna 430D	X		X
156	230	386	Cuyuna 215R	X		
220	240	460	Cuyuna 430D		X	
180	260	440	Cuyuna 215R		X	
185	240	425	Zenoah		X	
220	265	485	2-Yamahas		X	
160	240	380	Yamaha	X		X
198	220	418	Cuyuna 430D		X	
160	220	380	Chotia		X	
155	275	430	Yamaha	X		X
170	280	450	Chrysler		X	
135	230	365	Solo		X	

Hard Wing

120	230	N/A	CGS Powerhawk	X		X
240	240	480	Cuyuna 430D		X	
165	225	380	Yamaha J1000		X	
153	200	353	Solo		X	
145	230	375	Pioneer		X	
185	300	485	Zenoah		X	
152	195	347	Chrysler		X	

plus the weight of a 165 pound pilot, and this sum is divided by the wing "Area." N/A means the data are not available.

3 POWER SYSTEMS

In the early days of hang gliding, the standard Rogallo hang glider had a low glide ratio of about 4 to 1. This meant that only short flights were possible, unless the pilot could find an extremely good lift wind or thermal close to the launch site, to loft him upward.

Soon, very lightweight, simple power systems were introduced in hang gliders, to offer the pilot a longer flight, particularly for times when lift winds were weak or absent. These rudimentary systems were not thought of as ordinary airplane engines—which provide takeoff and flight thrust—but simply as "glide extenders," as they were often called. Although they might be left running throughout a flight to "extend" glides and assist in improving soaring, their fuel capacity was too small to make long flights practical, if the engine was left on from takeoff to landing.

Many believed, the best way to use these power systems was in taking off from level ground until reaching good lift winds. After this, the pilot could turn off the engine to conserve gas, then restart when or if the lift became too weak, fly on to find another strong lift wind or thermal and then turn the engine off again.

Back-Pack Power

Some of the first efforts to power hang gliders occurred early in the 1970s, when the sport was in its infancy. Among the first of the early power systems was Bill Bennett's McCulloch 101 engine, a ducted back-

Jerzy Kolecki of Stockholm, Sweden, designer and manufacturer, demonstrates his 30-pound, ducted-fan, *Motolotnia* (Polish for motorized hang glider) back pack power system. (See chapter opener photo also.) It uses the Chrysler 820 engine. *Photos by Jerzy Kolecki*

pack fan that looked like a 20-inch window fan suspended from the shoulders with web straps (see photos, page 42 and above). Humorously called "back massagers," others of similar design also appeared.

A writer, Noel Whittall, in the March 1981 issue of *Whole Air Magazine*, gave a colorful opinion: "These machines required a particular blend of courage and stoicism to operate, as although the influence on the glide angle was almost imperceptible, the physical effect on the pilot was quite marked. Reflect a moment: You have a two-cycle engine strapped between your shoulder blades, driving an imperfectly balanced propeller. When the engine starts, you need a pronounced lift to port to counteract the torque, plus a backward lean to balance the slight thrust. So far so good. Now pick up your hang glider and perform a controlled launch."

Fortunately, around this period it was discovered that a couple of dollars worth of plastic battens would extend the glide rather more effectively than a vibrating engine clamped on your back.

Bill Bennett said in the same issue of *Whole Air Magazine*, "The backpack engine was just another idea that we've had here that's been three or four years ahead of its time, like Mylar® leading edges, double-surface sails, and floating crossbars."

"The hang-gliding movement," he continued, "wasn't ready for power, and the gliders weren't ready for power. I don't believe any of those old standards [Rogallo hang gliders] would fly with any of the power packs we have available today."

It was only a matter of months after the backpack system appeared that the Soarmaster system was introduced. Its engine and long propeller shaft could be clamped to the keel of the glider. While it offered a more efficient application of power than the backpack unit, it was only moderately successful. Although its use did diminish in the mid- and late 1970s, it has recently had a strong rebirth, both under its original market name and in many adaptations. Not only had the application of power in its early forms occurred too soon in the development of hang-glider designs, but the systems were difficult to use. The pilot of the Soarmaster, for example, held the throttle control between his teeth. It was awkward to grasp and pull starting cords if the engine stopped in flight.

Although Whittall is correct that "plastic battens would extend the glide," there were soon many other improvements in design. They are described in *Hang Gliding and Soaring* by James E. Mrazek (published by St. Martin's Press, New York). Consequently a dramatic leap in glide ratios resulted, to the point that many models boasted 10-to-1 ratios. As the glide ratio improved, pilots found less and less need for "glide-extending" power attachments.

Recent Developments

Soon, major changes occurred in ultralights. Engines became larger, resulting in increased power. Net weights went up 70 percent. While early ultralights, particularly the foot-launched flex wings, weighed about 100 pounds, models were built close to 200 pounds, with some weighing as much as 300. Weight-shift models gave way to three-axis control models. Pilots were offered more instruments, and some models had enclosed cockpits. Ultralights flew faster and climbed more rapidly—and they became more expensive.

Tractor Versus Pusher Power Systems

Many manufacturers have changed to pusher engines because they put the prop blast behind the pilot and direct the airstream at the tail surfaces. This latter feature improves the control responses. The inherent disadvantage is that pilot weight must counterbalance engine weight in order to maintain a proper center of gravity. Depending on the weight of the pilot, the CG shifts, upsetting the craft's trim. Several manufacturers have designed new drive systems aimed at overcoming this fault. In the *Mirage, Quicksilver MX, Vector 600,* and *Rotec Rally*, the engine is moved forward, either ahead of the wing or at the midwing position, with the prop behind the wing on a long driveshaft. This maintains the advantages of pusher drive, with the CG less sensitive to pilots' weight differences.

Advances in Australia

While Americans and Canadians were struggling with getting the proper combinations of power systems and hang gliders, and their other ultralight designs were hardly in the experimental stage, a surprising advance had already occurred in Australia. In 1974, Ron Wheeler of Sydney built a revolutionary, high-aspect ratio, flexible-wing hang glider called the *Tweetie*. He soon added a power system, transforming it into a true miniature airplane which he named the *Skycraft Scout*. He was soon flying it extensively, three years before comparable models were being flown in America. By 1976, the Australian Department of Transportation, pressed by Wheeler, published an Air Navigation Order governing the operation of aircraft such as the *Scout,* and as Gary Kimberley remarked in his magazine *MiniPlane,* "it enabled the development of an entirely new branch of aviation and brought about the birth of an exciting new sport."

Development of Power Systems

Many of the early ultralight engines were modified twin-cylinder Cuyuna or other snowmobile and Pioneer and Partner powersaw engines, since these were the only small, lightweight motors available that provide sufficient power for flight. Now the industry has engines specifically designed for ultralights, but U.S. and overseas manufacturers moved slowly in developing and using these power systems for several reasons. One reason was the uncertainty as to whether the market would remain strong or taper off after a few hang-glider enthusiasts who wanted to fly powered planes had been satisfied. Another reason was the expense of designing and building an engine specifically for ultralight use. Well-known engine manufacturers in the United States were also concerned about legal ramifications, leading some to build major components while allowing ultralight manufacturers to assemble and complete the engines, using their own brand names. Now, however, engine manufacturers are becoming more actively involved in designing and producing their own ultralight engines.

Let us take a look at ultralight engines and their propellers, the keys to transmitting engine power into a useful force.

Engine Facts

Most ultralight engines are two-cycle, internal combustion engines with one or more cylinders. Stated simply, these engines convert thermal energy (created by igniting a gasoline-air mixture drawn into their cylinders) into power, which is transmitted by mechanical means to the propeller.

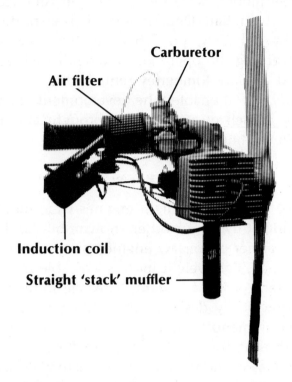

Twenty-five horsepower Chotia 460, a practical, simple engine, designed by Weedhopper especially for the *Weedhopper* ultralight. *Photo by the authors*

Two-Cycle Engines

Two-cycle means that the fuel mixture in the cylinder is ignited every time the piston is at the top of its stroke, making a power pulse for every revolution of the crank shaft. As the fuel and air mixture burns and expands, it pushes the piston down in the cylinder. About half-way down the cylinder this piston uncovers ports in its wall, allowing the exhaust to escape and be replaced by a fresh fuel and air mixture. In contrast, a four-cycle engine fires only every other revolution as is typical in a four-cycle lawn mower engine.

The two-cycle engine is simpler than the four, lighter in weight, and produces more power per engine weight. However, the two-cycle uses more gas than the four of comparable power, and its spark plugs are likely to foul sooner. The two-cycle engine is also noisier. The major advantages to the two-cycle engine are its simplicity of operation, its few parts, and its light weight.

This is a profile of a typical ultralight engine:

Power	10 to 30 horsepower at 5,500 to 10,000 revolutions of the drive shaft per minute
Weight	30 to 60 pounds
Displacement	100 to 500 cubic centimeters
Power/Weight Ratio	.65 to 1.20 brake horsepower per pound
Fuel/Oil Mixture	20 to 40 parts of auto gas to one part of oil, commonly written as 20:1, or 40:1
Fuel Consumption	1 to 2 gallons per hour
Cost	$500 to $1,500

In this chapter some of the features and operating characteristics of a typical engine are discussed. However, an owner/operator's manual is always an excellent reference for an engine. It should be strictly followed for proper operation and maintenance of the engine. Both two- and four-cycle engines perform well with proper operation and care and will last a long time.

Cooling

Engines are normally air cooled by air rushing from the propeller "propwash" through the fins of the engine, or by the motion of the craft through the air. Heat generated by the engine is initially dissipated to fins on the engine head, barrel, and crankcase, which then transfer this heat to the atmosphere. Some engines have axial fans in the engine housing that aid in heat dissipation. Proper cooling in these engines is important, since higher-than-normal operating temperatures can make cylinder heads overheat and may lead to engine failure.

A new engine among ultralight power systems, first introduced at the Experimental Aircraft Association (EAA) Sun 'n Fun Fly-in at Lakeland, Florida in March

1982, is the water-cooled *Spitfire*. Despite the weight added by the radiator and water jacket, the 35 horsepower unit weighs only 43 pounds. It is manually started with optional electrical starting. The water cooling system in this engine has some distinct advantages, supplying the carburetor with heat to prevent icing during cold temperatures, and, with proper ducting, also keeping the pilot warm. The manufacturer boasts low fuel consumption, reduced noise, and less vibration than many standard air-cooled engines.

Design

Ultralight engines are either, single, two (twin), or three cylinders. Single-cylinder engines such as the Japanese Yamaha or Zenoah, are small and have one main head. Two-cylinder engines are of two designs. One is "in line" with one cylinder behind the other such as the Cuyuna 430D; the other is "opposed" with one cylinder opposite the other such as the Italian KFM 107E. A three-cylinder radial engine now in use primarily in Canada is the German Koenig.

Some pilots favor twin engines like those found on the *Lazair*, the *Kolb Flyer*, and the *Hummingbird*. Advantages claimed for twins are a more uniform distribution of weight and airflow, no prop blast in the pilot's face as is true with front prop mounted single engine planes, and most importantly, backup power if one engine quits. One problem that crops up unless the engines are closely synchronized is the tendency of the craft to yaw because of unequal thrust between the engines.

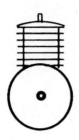
Single cylinder

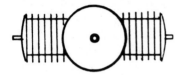
Twin cylinder (opposing)

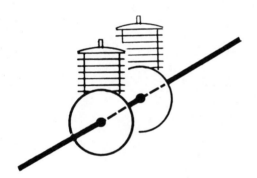
Twin cylinder (in line)

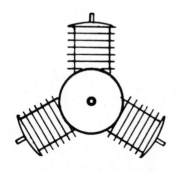
Three cylinder (radical)

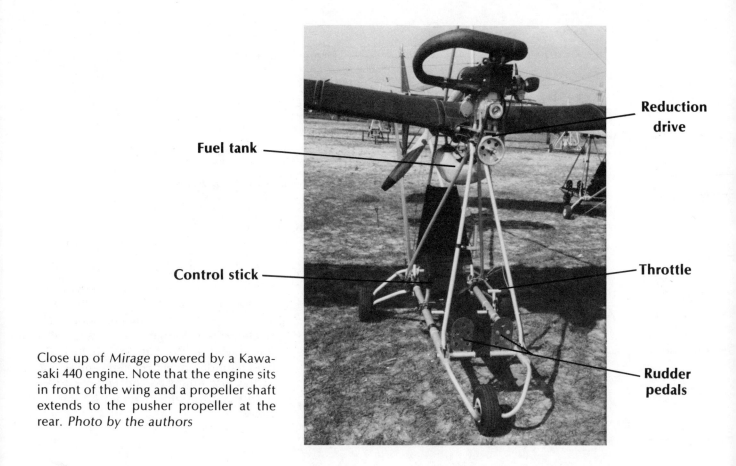

Close up of *Mirage* powered by a Kawasaki 440 engine. Note that the engine sits in front of the wing and a propeller shaft extends to the pusher propeller at the rear. *Photo by the authors*

Kawasaki 440, 40-horsepower, twin-cylinder (in line), air-cooled, reduction-drive engine. This is a pusher (propeller in rear) configuration. *Photo by the authors*

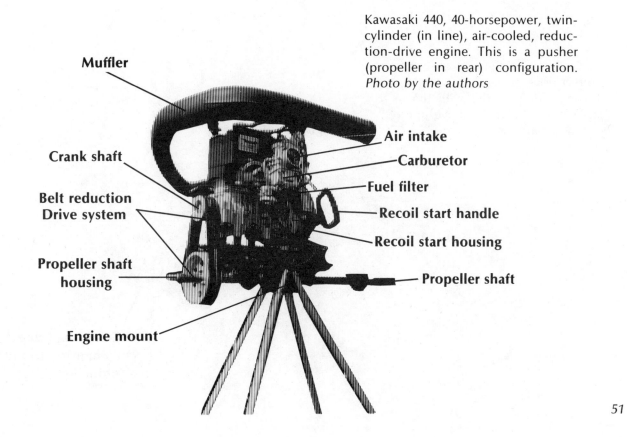

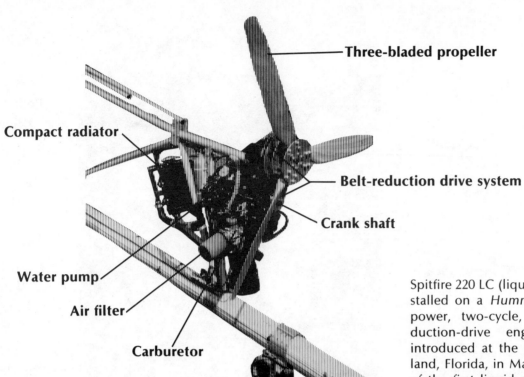

Spitfire 220 LC (liquid cooled) engine installed on a *Hummer*. It is a 30-horsepower, two-cycle, single-cylinder, reduction-drive engine. It was first introduced at the Sun 'n Fun in Lakeland, Florida, in March 1982, and is one of the first liquid cooled engines to be used by an ultralight. *Photo by the authors*

Thirty-horsepower, Cyuna 430D engine in a direct-drive configuration used in a *Goldwing*. Compare this with the reduction drive shown in above photo. *Photo by the authors*

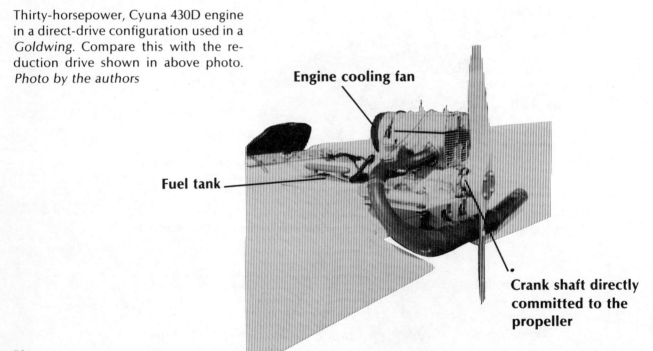

The twin cylinder (opposed) 25 horsepower Austrian Limbach engine. *Photo by Ed Sweeney, courtesy of Gemini International*

30-horsepower two-cycle, twin-cylinder (in line) Cuyuna 430D. It is air-cooled and has a reduction drive. It is used as a pusher-power system for this ultralight (engine up front, propeller in the rear). *Photo by the authors*

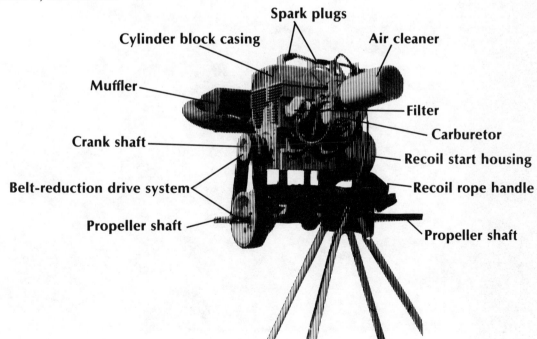

Mid-wing mounted Cuyuna 430D (installed on a *Hi-Nuski* ultralight). It is a direct-drive pusher. Note the bearing block adjacent to the propeller to support the extension of the propeller shaft. *Photo by the authors*

A "one-of-a-kind" power system using a direct-drive rotary engine to power a *Weedhopper*. A simulated machine gun is below the wing in front of the pilot seat. *Photo by the authors*

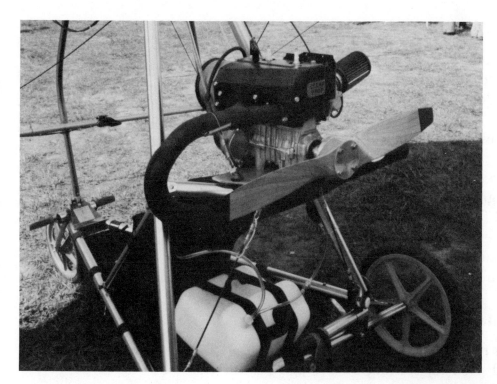

A direct-drive Kawasaki 400 powers a *Jet Wing* trike. *Photo by the authors*

Installed in an *Eagle*, a 20-horsepower Zenoah 250, two-cycle, air-cooled engine in a pusher configuration. *Photo by Dan Johnson, courtesy of* Whole Air

Fuel

Two-cycle engines require a fuel-oil mixture for internal lubrication and assistance in cooling, since they do not have an independent oil system as automobile engines do. When the engines are new, break-in fuel mixtures are normally lowered, such as from twenty-five parts of fuel to one part of oil, to fifteen parts of fuel to one part of oil. Break-in times vary with the engines, therefore, if an engine is new, be certain to consult the owner's manual for the proper mixtures. Normally, fuel for ultralight engines is regular automobile gas. Premium or regular and a minimum octane rating may be specified in the manual for the engine.

Starters

An ultralight engine is normally started by recoil, by much the same method used for starting an average lawnmower. There are some engines that offer self-starters, but presently the weight of this starter and of the battery required to run it makes their use impractical. Starting is not difficult; some ultralights can even be restarted while in flight after the engine has been shut down to take advantage of engineless gliding on thermals.

Only under certain conditions should an engine be started by hand cranking the propeller: if the recoil starter or other normal mechanism is broken *and* the *engine manufacturer recommends it* as a safe procedure; or if it is the manufacturer's recommended way of starting an engine. It is important *not* to start the engine by this method unless it is recommended because an ultralight engine turns the propeller at thousands of revolutions per minute, and once the engine fires and the propeller turns, it moves *very* quickly. This is no time to experiment or take chances. More about this subject is discussed in Chapter 7.

Reliability

Reliability is a term heard often in aviation circles. It applies mainly to engines, but also applies to altimeters and other instruments and parts of the ultralight.

When speaking of engines, reliability refers to how well the machine keeps running without sputtering, missing, losing power, or stopping.

More than any person using an engine in any other vehicle, a pilot especially cannot afford to have an engine that may give trouble or stop while he or she is airborne. In an effort to obtain that reliability, components have been strengthened and improved in some engines. Other engines have been "detuned"; that is, they have had their compression ratios lowered and other changes have been made so that the engine operates below its maximum output. This increases engine life and reliability because engine wear and stress are reduced. Two-cylinder engines offer the added insurance that if one cylinder sparkplug should quit, the engine will still operate, although at a reduced power. For the most part, ultralight engines are reliable when given the proper care and maintenance.

Propeller (Prop)

Design and manufacture of propellers is an art. A propeller has the same aerodynamic objectives as a wing. The wing produces lift, the propeller thrust. Each uses an airfoil-shaped cross-section. On close examination, it is possible to see that one side of a propeller is rounded and the other side is flat. Propellers vary in shapes and sizes, depending on the effects desired during flight. Propellers with a greater pitch perform better during the cruise phase of flight, although they do not perform as well during takeoff. The designs of most propellers used on ultralight aircraft represent compromise between two needs: cruising flight versus takeoff demands. However, it is at cruising speeds that propellers usually operate most efficiently.

Variable-pitch propellers—those that can be rotated at the hub to change pitch during various phases of flight—are more efficient and are used on many conventional aircraft engines. Someday they may be used with ultralights, but are now only in an experimental stage. Normally, two-bladed props are used today, although three-bladed props may well be adapted to use on ultralights in the future. An engine with twin-counter rotating propellers is also available for ultralights. These propellers are on concentric shafts; each shaft rotates in opposite directions.

Ultralight engines usually come equipped with a propeller. If not, the manufacturer recommends several propeller designs that can be used. Unless a pilot is an engineer and well acquainted with engines and propellers, it is best to follow the manufacturer's recommendations. The manufacturer knows which combination or combinations of engines and props are the best for the safe and efficient operation of the combination of a particular engine and ultralight.

Direct Drive and Reduction Drive

The propeller is attached to the engine in two different ways. In one, it is attached directly to the propeller shaft, rotating once with every rotation of the engine. This is called direct drive. The other, reduction drive, uses a device to lower the speed of the propeller in relation to the engine's speed. Reduction units are commonly used in ultralight engines.

From an engineering standpoint, there are several good reasons for using reduction drive. Reduction units are generally added to engines used in low-speed aircraft to produce greater engine efficiency and less engine wear, because longer propellers can be used that rotate more slowly. Since the propeller tip cuts through the air more slowly, it makes less noise than a propeller with direct drive. Moving prop tips create most of the noise that comes from airplanes, and the slower they move, the less noise is generated altogether. Reduction drive units use chains, gears, and rubber belts. The V-belt reduction units use two to three belts in ratios where the revolution of the engine shaft turns two (technically shown as 2:1) or more times for each one revolution of the propeller. The majority of reduction systems in ultralights are 2:1 reduction systems although some go as high as 6:1.

Mufflers

In addition to the noise created by the propeller, the exhaust noise of two-cycle engines is considerable.

With the engine so near the pilot, noise can deafen him. It may also disturb people below or nearby. Some engines have no muffler(s), only a short porting that directs the exhaust away from critical areas. Others have larger mufflers that cause a slight decrease in available power. Yet other mufflers are available, specifically designed and acoustically "tuned" by their shape and size, to cause no loss in power and, in some cases, to actually boost power.

Maintenance

Maintenance of an ultralight power system is relatively easy. With good care, and observance of a system's operating limitations to ensure it is not abused, it is possible to obtain hundreds of hours of trouble-free service. Some manufacturers now claim a possible one thousand hours of operation or more before a complete overhaul is required, in which the engine is taken apart and parts cleaned and replaced as needed.

Keeping the sparkplug(s) clean and properly gapped, the carburetor properly adjusted, the fuel-oil mix correct, and doing other minor tasks as specified by the owner's manual will help keep an engine and propeller in good shape. Consult the owner's manual accompanying the engine for guidance on these matters.

Future Prospects

In the next several years, as momentum in the ultralight field grows, engines will become lighter in weight, with more power, and more reliability even as ultralights themselves become heavier, requiring more power. More exciting is the prospect that in the future the ultralight power system may not be a traditional internal combustion engine. For example, propulsion took an innovative advance when Ted Aucona of Hollywood, California, built and flew a jet powered *Icarus VI* hang glider in mid-1981. Developed by Eugene Gluhareff, the engine weighs 24.5 pounds, has no moving parts, and uses liquid propane as fuel.

Some farsighted people are doing fine work with solar-powered ultralights, such as the *Solar Challenger*,

ENGINE TECHNICAL DATA*

ENGINE NAME	Displacement (cubic centimeters)	Weight (pounds)	Horsepower (at RPM)
Chotia 460B	456	31.5	25 at 3800
CGS Powerhawk 152	380	59	20 at 5500
Chrysler 82026	135	13.5	10 at 8000
Cuyuna 215R	214	3.9	20 at 5500
Cuyuna 430D	428	64	30 at 5500
Fuchs	N/A	N/A	N/A
Gemini/Partner K 1200	100	13	8 at 2500
Gemini/Limbach 275	275	14	16.7 at 4800
			25 at 7500
Kawasaki 440	440	55	38 at 6000
KFM 107E	294	42	25 at 6300
Koenig	430	34.5	18 at 4500
Lloyd 386	386	N/A	22 at 6500
McCulloh MC 101	123	12.3	12.5 at 9000
Rotax 300	300	42	25 at 6000
Skylark	320	35	22 at 4800
Spitfire 220LC (Liquid cooled)	220	43	35 at 8500
Yamaha KT 100S	100	23	15 at 10000
Zenoah	250	17.5	20 at 6500

* This index is a partial listing of the many systems available. Specifications are supplied by the manufacturers. Although listed, it appears that the Chrysler, McCulloh, and Koenig are waning in popularity.

designed and built by a team led by Dr. Paul MacCready. New very lightweight batteries of much stronger power and duration than we now know are being studied and may lead to the development of an ultralight with an electric engine and no solar cells, or perhaps a combination of both. (For more information on this development see Chapter 12.)

Current propeller design could completely change in the next ten years. Civilian manufacturers and the Na-

Cylinders	Cylinder Bore (inches)	Piston Stroke (inches)	Fuel Oil Mix (after break in)	Fuel Consumption (gallons per hour)
1	3.46	2.95	40 to 1	1.5
2	2.037	2.44	40 to 1	1.3
1	2.53	1.62	20 to 1	1.0
1	2.66	2.36	40 to 1	N/A
2	2.65	2.36	40 to 1	2.5 to 2.0
N/A	N/A	N/A	N/A	N/A
1	2.20	1.58	48 to 1	.4
2	2.6	1.57	50 to 1	.75
2	2.67	2.36	25 to 1	2.0
2	2.40	2.10	N/A	1.8
3	N/A	N/A	50 to 1	1.5
2	N/A	N/A	N/A	N/A
1	2.28	1.83	20 to 1	1.0
2	N/A	N/A	50 to 1	1.25
2	3.24	2.12	25 to 1	N.A.
1	2.7	2.35	24 to 1	1.3
1	2.08	1.84	25 to 1	1.5
1	2.83	2.34	25 to 1	1.0

tional Aeronautics and Space Administration (NASA) are today testing "propellers" that look like the head of a bullet in their radical sweep and pitch. These new propellers promise much more efficient performance along with lowered noise levels. New, extremely lightweight, advanced technology materials called "composites" may someday replace wood and metal, not only in propellers but in major structural parts.

4 CONTROL SYSTEMS

Three different control systems are used in ultralights: weight-shift, weight-shift supplemented by one- or two-axis aerodynamic controls, and three-axis, complete aerodynamic control. Trends indicate that two- and three-axis-stick-control-system ultralights are becoming more popular than those combining the weight-shift system. However, all systems are designed by the manufacturers to provide adequate control authority.

Weight Shift Control

This is an uncomplicated system; the pilot simply shifts his body to control the direction of flight. A change in the position of the body weight moves the craft's center of gravity, tipping the ultralight into the direction the pilot wishes it to fly. The pilot moves his weight forward to gain airspeed or lose altitude, and backward to increase the angle of attack and gain altitude. He moves his body along the control bar to the right to bank to the right, and left to bank to the left. This system most nearly parallels the method of control used for most hang gliders and is familiar to those who have flown them. In ultralights it is generally used in combination with other aerodynamic controls.

Many pilots feel there are disadvantages to the weight-shift-control system. They claim that the craft responds sluggishly in comparison to two- or three-axis control systems with standard control elevators, ailerons, and rudders. They also claim the inability to easily separate roll and yaw.

Weight-Shift with Supplemental Aerodynamic Control(s)

In this system, pitch control is by weight shift, and in addition, there is some kind of rudder control or drag device. Pitch control in some craft is governed by a combination of weight shift and an elevator. The rudder and spoilers can be controlled by a foot device or hand controls. This gives the pilot a greater degree of control of his craft without adding too much weight. However, these additions add the problems associated with mechanical devices and cables, such as an increased chance of the failure of a mechanical part, more maintenance, added weight, and increased drag.

Three-Axis (Stick) Control

Aerodynamic three-axis control, sometimes called "total aerodynamic control," is described in Chapter 6. With this method, there are separate controls to move the ultralight about either the vertical, longitudinal, or horizontal axes. Stick or rudder pedals activate a rudder, ailerons, and elevator(s) to cause these movements. Some designs improve control with spoilerons, tip rudders, and canards with elevators. All are familiar to the conventional pilot. These additional control devices have their disadvantages in that they force the pilot to learn about more systems. While they help to make the craft fly better, they make the flying slightly more complex.

Inverted "V" tails are aerodynamically very efficient tail designs. The reason this tail is not used more often is structural; the problem being that the wing tips of the tail are close to the ground and subject to obstruction, and also, with the tips, there is a rotation problem for slow steep ascent takeoffs.

In some designs this has been overcome by cable bracing each tail to the forward fuselage and by setting the wing to boom incidence at 10°, so that we have a tail dragger aeroplane that can be rotated to takeoff and land.

The Best System

A good way to evaluate and compare these systems is to talk to the people who have had them on ultralights they have flown and take into consideration your own experience. Other pilots can tell a lot about how the controls of a particular ultralight handle. Sensitive "seat-of-the-pants" flyers or hang-glider pilots may prefer weight-shift controls. On the other hand, a licensed aircraft pilot, used to a three-axis control, may be more comfortable flying an ultralight controlled the same way.

Canards

"Tail first" canards are found in the *Goldwing, Eagle, Tomcat,* and the *Pterodactyl Traveler,* and act as stabilizers and are also used for pitch control. Canards stall first as speed decreases, dropping the nose automatically and restoring speed and lift before the main wing can stall, thus theoretically making canard ultralights stall- and spinproof resistant.

5 INSTRUMENTS AND GEAR

Ultralight flyers use varying amounts of instruments depending on their needs. Many pilots who do local recreational flying at low altitudes find they do not need instruments. A pilot who rarely flies above 200 feet really does not need an altimeter. One who does no engine-off soaring does not need a variometer. However, now that ultralights are becoming more versatile, and are approaching light airplane performance standards, instruments such as airspeed indicators, altimeters, variometers, and tachometers are being used increasingly (see photo, page 66). Pilots find that instruments tune them into the ultralight and also enable them to exploit their craft and the air to a greater extent.

Airspeed Indicator

The airspeed indicator registers the forward progress of the aircraft in relation to the air through which it flies. If that air is still and the ultralight is flying at 40 miles per hour, it progresses through that still air at 40 miles per hour (across the ground at that speed also).

However, if the wind is moving at 20 miles per hour in the direction of flight, and the craft is cruising at 40 miles per hour, the ultralight is moving 60 miles per hour in relation to the ground, but the air speed indicator will still read 40-miles-per-hour airspeed. If the ultralight is flying against that same 20-mile-per-hour

Instrument panel of a *Hummer.* Note the pull chord handle at right. Strap holds control stick forward. This panel has the usual instruments found in a light airplane indicating that this *Hummer* is used for many kinds of flying. *Photo by authors*

wind it is moving forward only at 20 miles per hour relative to the ground, and the indicator continues to read 40-miles-per-hour airspeed. This is why takeoffs and landings are made into the wind, so as to have the most airspeed for the least amount of ground speed. The *ultralight reacts to airspeed,* not ground speed. Only at the appropriate airspeed, which will always be higher than the rated stall speed of the craft, can the pilot maintain control.

On the airspeed instrument, the indicator hand moves over the face of the instrument and registers the speed in miles (or kilometers, or both) per hour. The hand is activated by the increasing, or decreasing, wind pressure in a tube, usually located at the forward edge of the wing where it is not affected by the propeller blast. The pressure recorded is proportional to the speed of the ultralight through the air mass.

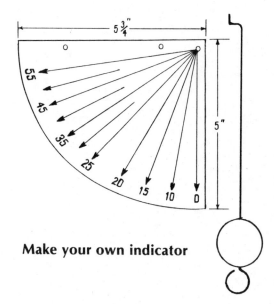

Make your own indicator

There are many airspeed indicators. Some are elaborate, expensive, and very accurate. Others are simple, but can suffice. One is a glass tube with a hole facing into the wind and a simple wafer that moves up and down with the pressure. The calibration at which the wafer settles on this see-through tube indicates the craft's airspeed.

A simple 3-ounce indicator can be made at home from a slice of Plexiglass, a piece of wire, and a Ping-Pong ball. It is the invention of Robert Hansen of Lake City, Florida. The illustration will help to duplicate his efforts.

Cut the Plexiglass to shape. Drill the pivot and mounting holes. Sand the edges of the Plexiglass and paint the backside white, if desired.

Insert a length of wire through the Ping-Pong ball as close to the centerline as possible and bend it as shown. When this has been done, it is ready to be calibrated. Pick a day when there is no wind. Hold the instrument as far out a car window as possible as someone else drives. At 5 miles per hour, mark the position of the end of the wire nearest the ball with a red grease pencil mark on the plexiglass. Speed up to 10 miles per hour and repeat the process. Do this for speeds up to 55 miles per hour. Obtain presstype arrows and numbers from an art store. Rub them off on the plexiglass at the appropriate red dots; erase the dots. As Joel Strait says in *Ultralight Flyer*, "... a good tach and this simple foolproof airspeed indicator will go a long way toward producing old, not so bold, ultralight pilots. This little gadget works!"

Incidentally, Bob sells them already calibrated for $9.95. His address is 799 E. Duval Street, Lake City, Florida 32055.

Altimeter

An altimeter tells how many feet the ultralight is above either sea level or a particular place on the earth for which it is set. It is a gauge that reads altitude by changes in pressure upon a diaphragm within the instrument. As the craft rises into thinner air where there is less pressure, the diaphragm expands, causing the hand of the altimeter to show a higher altitude.

An altimeter must be adjusted for local barometric readings. The pilot learns the local barometric pressure by calling an airport weather service or, more often, someone in the air-traffic-control tower, and adjusts the altimeter to correspond with the correct local barometric reading before takeoff. Altimeters give readings in feet, and many are accurate within plus or minus 10 feet of actual altitude.

Variometer

Sometimes referred to as a climb indicator, also as a vertical speed indicator, the variometer tells the pilot his rate of change of altitude measured in feet per minute. As with the altimeter, this change takes place when the barometric pressure changes. It is sensitive to the most subtle changes in lift, giving immediate clues to the presence of an unseen thermal or other lifting air force.

Although the basic function upon which all variometers are designed is the indication of change in barometric pressure, there are a number of distinctly different variometers manufactured. One variety has a dial face similar to that of the altimeter with the hand moving into either a plus or minus area indicating the rate of climb in feet. Another instrument warbles in rising air. This feature would be of particular value to an ultralight pilot who shuts off his engine to soar. The point at which the warble tone starts is adjustable and can be set by the pilot prior to launch. The pitch of the continuous tone is directly proportional to the rate at which the glider is rising or sinking, and the pilot can detect a change of 30 feet per minute or less. As the pilot becomes accustomed to flying with the instrument, he comes to react automatically to the changing tone. Response time of the tone in the instrument is usually a fraction of a second.

The pellet variometer is the most commonly used, especially with hang gliders and in sailplanes. Highly sensitive, it is the favorite of soaring pilots, and is also relatively inexpensive and reliable. Its design is simple. Its action is equally uncomplicated.

As the sailplane goes lower, the air pressure (atmospheric pressure) outside gets greater and forces air into the outside opening of the variometer tube. This, in turn, causes the several red balls in the left tube to rise and the green ones to stay at the bottom or descend to the bottom and seal the tube at that point.

If the ultralight gains altitude, the pressure within the system, now greater than on the outside, causes a reverse action of the balls. The green ones go up in their tube. The red ones go down to seal their tube.

By closely watching the instrument's action, the pilot can tell when he is going up or down, and at what rates. The higher the rate of ascent, the higher the balls will rise in the tube, dropping gradually as the rate of ascent slows down. Beginning pilots find the action of this clever instrument one of the most absorbing phenomena on the instrument panel.

Tachometer

Experienced ultralight flyers favor including a tachometer among their instruments. It registers the revolutions per minute (rpm) that the engine driveshaft or propeller shaft is making. A glance at it tells the pilot if the engine is operating at the proper idle speed or faster than it is designed to go. A tachometer can alert the pilot that the engine is in danger of malfunctioning from too many revolutions. The pilot brings the engine to proper performance by adjusting the timing or the carburetor or increasing or decreasing throttle control. The tachometer is an indicator of this performance.

Cylinder Head Temperature (CHT) Gauge

It is important to know the temperature of the engine, as higher-than-normal temperatures can lead to many problems which cause engine failure. Temperature is measured at the cylinder head, a critical component of the engine, and then transmitted by wire to the CHT gauge. If operated within temperature ranges specified in the owner's manual, the life of the engine will be prolonged.

Radio

Several manufacturers now produce two-way radios especially designed for ultralight flying. One costs about ninety dollars, weighs 5 ounces, can be attached to either helmet or harness, and has a range of 3 to 5 miles. Such radios are valuable for instruction, enabling the instructor to talk the pilot into correct flying modes or answering the pilot's questions. Some instructors launch behind their students and follow them in flight, observing their flying techniques and directing corrections over the radio. UHF/FM radios operate effectively, little affected by ignition noise. Do not buy a CB hookup. CB is AM, and therefore useless because of the engine ignition interference.

Helmets

Accidents will happen. The pilot's head is the part of the body most likely to be injured in a crunch, either by hitting the ground or hitting wires and tubes. A helmet designed for flying will absorb impact and lessen the degree of injury, if not prevent injury altogether in most accidents.

Although it would seem that motorcycle, football, and other sports helmets would meet the needs of the pilot, this is not so. Some of these helmets are excessively heavy. Most helmets used in other sports lack adequate ear exposure, keeping the pilot from feeling or hearing the wind or eccentric sounds from the engine, struts, and wings that can be critically important.

Thus, it is recommended that the helmet bought be especially designed for ultralight flying. It should be comfortable, penetration resistant, lined with energy-absorbing foam, lightweight, and not impair the vision or hearing. The price of a quality helmet is about $35.

Fabric Care

Most modern fabrics used for ultralight wings and other surfaces are Dacron or other polyesters. They should be kept clean, dry, and trim.

Dirt and sand will tend to abrade the fibers in the sailcloth, causing it to age prematurely. If the cloth is constantly being set up or disassembled on sandy or dusty soil, it's a good idea to hose it down periodically with fresh water and let it dry in the shade. If that doesn't remove the dirt, take the sail off the frame and wash it in mild detergent or soap.

Detergent and sunlight are a destructive combination. If the ultralight must be washed, rinse it well. Then rinse it once more. Get every trace of detergent or soap out. Then let it dry in the shade, not in direct sunlight.

Clean small spots of oil or grease with alcohol, a cleaning fluid such as perchlorethylene, or one of the various sail cleaners on the market, but don't expect dazzling results. Some say stains respond to Clorox bleach, applied cold, full strength, and followed by a thorough rinse. This treatment is especially recommended for mildew stains.

The best weapon against mildew, of course, is prevention. Let the sail dry out in the shade before rolling it up. Never put an ultralight that is damp or wet into storage. Always make certain it is completely dry.

6 THEORY OF FLIGHT

Before flying an ultralight, wisdom suggests that the budding pilot learn the fundamentals of flight. However, the present array of "unconventional" designs with canards, tip rudders, inverted V-tails and the like sometimes defeat anyone's attempt to understand how these aircraft fly. Yet there are certain fundamentals common to them all. This chapter will not make an aeronautical engineer of the reader. (A local library is an excellent source of additional information should the reader want to learn more.)

The Ultralight in Flight

There are four forces acting on any aircraft in flight (see illustrations pages 73–74):

- Lift   The upward-acting force generated by the airflow over the wing (explained further below).
- Weight   The downward force exerted on the aircraft's mass due to gravity.
- Thrust   The force developed by the propeller to give the ultralight its forward motion.
- Drag   There are two kinds of drag, parasitic and induced. Parasitic drag is the re-

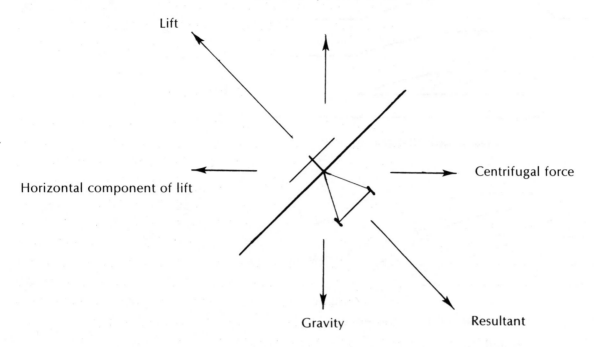

Forces acting during a turn

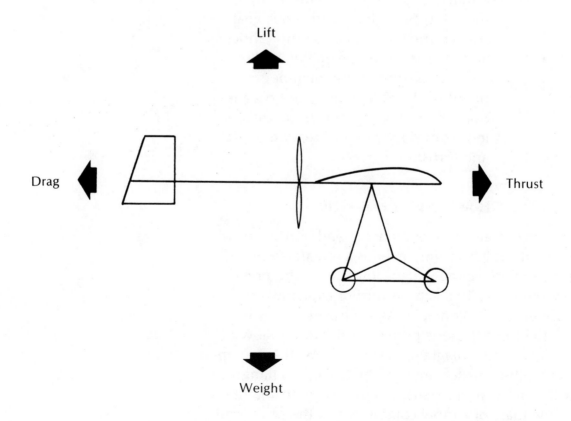

Force components

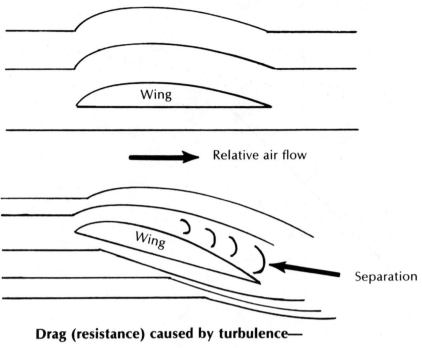

**Drag (resistance) caused by turbulence—
A stall in the making**

tarding force caused by the body of the pilot, his helmet and other gear, surface friction caused by the fabric, and the friction of the engine and the wires and supporting members exposed to the airflow. Induced drag is caused by slowing effects of turbulence developed in the creation of lift and thrust.

How a Wing Creates Lift

Efficient ultralight operation depends on a knowledge of lift, which, in simple terms, is created when the air pressure above the wing is lower than the pressure below the wing. This can be further explained by the use of Bernoulli's Principle: As airflow increases in velocity, the force it exerts perpendicular to the flow decreases (see top diagram, page 75). As the air approaches the smaller area of this tube, a change in velocity, pressure, or density must occur to maintain the same mass of airflow coming out of the other end.

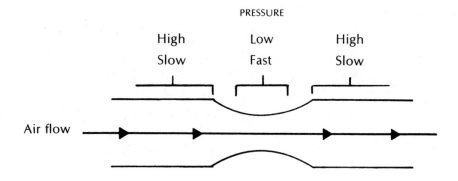

How lift develops (1)

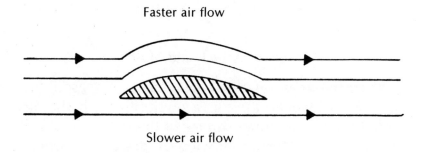

How lift develops (2)

Subsonic airflows usually change in pressure and velocity, and as velocity increases, there is decreased static pressure. What does this have to do with a wing? Notice that the area in the tube that is narrower, where the air moves faster, resembles the top part of a wing in its curvature (see diagram above).

The camber of an upper surface is different than that of the lower surface, which is usually flat. With the upper wing surface the same as the middle of the tube, pressure on the top part of the wing will decrease from the increased velocity of the airflow: The airflow below the wing is slower than that above. The difference in the pressure between the top and bottom of the wing creates lift.

Some terms frequently used to describe parts of the wing are:

- *Leading Edge,* or front of the wing.
- *Trailing Edge,* the opposite end from the leading edge.
- *Camber,* the curvature of the wing's upper and lower surface and normally, as explained in the creation of lift, the upper surface is more curved than the lower surface.
- *Relative Airflow,* the airflow generated by the movement of the airplane. Airflow is always opposite the flight path of the aircraft and is equal to the speed of the craft.
- *Chord Line,* an imaginary straight line between the leading and trailing edge.
- *Angle of Attack,* the angle between the wing chord line and the relative airflow.

Angle of Attack and Stalls

When the angle of attack or pitch is increased, lift increases. However, there is a point at which the air cannot follow the upper surface of the wing and starts to separate from the upper-wing surface and "burble" (see diagram below). At this point, because the airflow is disturbed, the wing loses its lift and "stalls"—and drag greatly increases because of excessive angle of attack, not because airspeed is too slow.

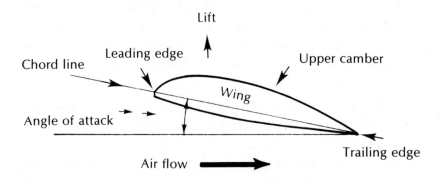

How lift develops (3)

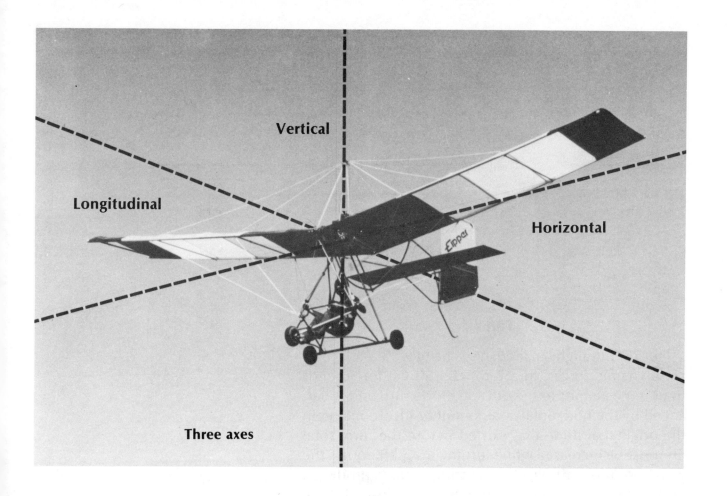

The Three Axes

An aircraft maneuver will involve one or more of the three axes upon which an aircraft moves. These are the longitudinal or roll axis, the lateral or pitch axis, and the vertical or yaw axis (see illustration above). Cockpit controls activated by the pilot with his hand(s) or feet move control surfaces and cause movement about the three axes. Maneuvers, however, are controlled in several different ways. An elevator controls pitch, rudders control yaw or movement about the vertical axis (and sometimes both yaw and roll in high dihedral ultralights such as the *Quicksilver*), and ailerons control roll. Body movement is also used for maneuvering in many ultralights; the shift of a pilot's weight moves the center of gravity of the craft, causing it to pitch, roll, dive, or climb. This procedure is described in greater detail in Chapter 4.

Control movements

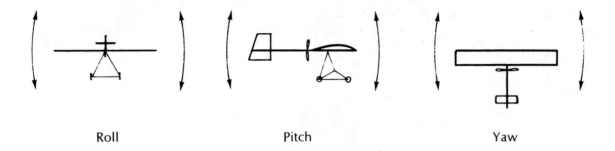

Roll   Pitch   Yaw

Turns

During ultralight maneuvers, changes take place in aerodynamic forces. In a turn, centrifugal force tends to pull the aircraft to the outside of the turn and is balanced by the horizontal component of lift. To maintain the original altitude that existed before the turn, total lift must be increased while turning until lift equals the weight of the craft plus the centrifugal force produced by the turn. An ultralight in a turn will lose altitude unless the angle of attack is increased to produce more lift. This increased angle of attack increases drag, so airspeed will be lost if power is not increased. As turns increase in bank, centrifugal forces increase, and so must the angle of attack increase. In a 60° banked turn, the lift requirement of the craft becomes twice what it was in level flight. Wing loading increases so the stall speed increases.

Glide Ratio

Once it is in the air, an aircraft battles the inevitable force of gravity. An aircraft's glide ratio is important when the engine is throttled back, turned off to enable the pilot to soar, or if the engine stops unplanned.

If an engine stops, gravity takes hold and inevitably pulls the craft toward the earth. It descends gradually a number of feet proportional to a certain number of feet forward, depending on the glide ratio designed

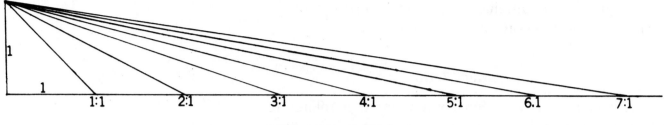

Glide ratios (L/D's)

into the craft. Glide ratio is a comparison of the feet of distance forward in still air to the loss of feet in altitude in that forward distance in the same period of time. For example, if the glide ratio is 8 to 1, the ultralight moves 8 feet forward for 1 foot down.

The glide ratio plays an important part in the performance of an ultralight. In essence, the higher the glide ratio is, the less energy (thrust) that is necessary to propel the craft through the air. When the engine is turned off, its ratio is higher and it will glide farther. Generally the higher the glide ratio, the lower the sink rate—the minimum rate of descent in still air—and the better the craft therefore responds to lifting thermals or slope or other winds when its power is off. This means that less power is needed to keep it aloft. Most ultralight glide ratios are better than 8 to 1. The *Lazair* ratio is 13 to 1 and the *Mitchell Wing* B-10, about 18 to 1.

Wing Loading

The area of the wing and the weight each square foot carries in flight determine wing loading. This is important in determining the lift characteristics of the craft. To determine the wing loading, take the total loaded weight of the craft, and the weight of the pilot, and divide by the square-foot area of the wing. For example, if the pilot weighs 165 pounds, the empty ultralight weighs 155 pounds, and its wing area is 160 square feet, the total weight in the air is pilot (165) plus craft (155), or 320 pounds. Dividing this 320 pounds by 160-square-foot wing area gives the wing loading, which, in this case, is 2 pounds per square foot. Lighter wing loadings decrease or allow lower flying speeds and generally mean a more docile handling machine although in such ultralights, they may respond rather sluggishly to the controls.

Aspect Ratio

Aspect ratio is a design component that relates to performance and is essentially a measure of the "skinniness" of the wing. The aspect ratio equals the wingspan divided by the wing chord—wingspan being the distance from wing tip to wing tip. For example, if the wing has a 30-foot span and a 5-foot chord, the aspect ratio is 30 divided by 5, which is 6. A wing with a short span and wide chord creates a lot of drag at low air speeds. Increasing the aspect ratio by increasing the length of the wing and narrowing its chord improves this efficiency. The more efficient the wing, the longer the glide. It makes sense, then, that designers make the wings longer and narrower, particularly for gliders and ultralights, in an effort to improve the low speed efficiency of the craft.

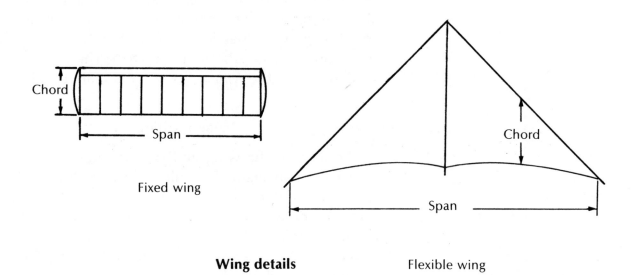

Wing details

7 LEARNING TO FLY THE ULTRALIGHT

Flying an ultralight is, in the estimation of those who have flown both hang gliders and airplanes, somewhat closer to flying a hang glider than flying an airplane. The similarities between ultralight- and private-pilots' training may increase as the federal and state governments raise requirements for ultralight operation.

What follows is not an instructional manual. It represents a synthesis of a growing body of knowledge and is general in its application. It is not a substitute for the instructional manual that accompanies each craft or for the skilled guidance of an instructor.

To excel in any sport or endeavor, it is necessary to take the time to prepare for it. This is especially true in flying. Once a person is at the controls and is airborne, there is little time to divert attention from the act of flying to work problems out. The ultralight is small and simply built. This makes many flyers overconfident and less cautious than they should be. It is necessary to take many careful, deliberate steps before the first flight. If taken, chances are the pilot will be around for his second flight.

Those who think this book talks too much about caution, safety and deliberate preparation should pick up a newspaper or magazine that deals with ultralights. In every issue those experienced in the sport or preconcerned readers advise those who are just starting ultralight flying to take extensive training.

Even those who are experienced airplane pilots are cautioned to train for ultralight flying. In *Ultralight*, Guy Kimberley of Blakehurst, Australia, cautions, "Previous light aircraft flying experience, whilst helpful, will not provide a pilot with the skills necessary for flying ultralights or minimum aircraft. These revolutionary new flying machines often require such radically different flying techniques that the conventional pilot can actually be at a disadvantage because of inappropriate, ingrained and instinctive reactions developed over many hours of conventional aircraft flying."

Ultralight Flying Programs— Where to Find Out about Them

The most direct method is to ask the manufacturer or dealer from whom you want to purchase an ultralight. He or she should have literature on programs and be able to answer any questions. It would be beneficial to ask other dealers about flying training programs they offer and then select the best one.

Other ways to find out about flying programs are to read of them in the *Glider Rider, Whole Air Magazine, Ultralight Flyer,* or other ultralight magazines. Listings and advertisements in these periodicals deal with flying programs. Check them out.

Manufacturers and Flying Programs

There are many different ultralight manufacturers and just as many approaches to learning how to fly the aircraft. The Professional Ultralight Manufacturers Association (PUMA) has set an objective of creating a uniform set of standards for flying training. This will take time to implement and enforce, however. In the meantime, three different training approaches have evolved:

1. The manufacturer sells directly to the individual and sends a kit containing the ultralight, which the pilot assembles. The kit may contain a flight instruction manual, Federal Aviation Regulations (FARs) and a manual about mete-

orology. Experience shows that there is a high percentage of accidents among the self taught. Instructions may also contain recommendations to visit with other ultralight pilots or the local airport to get several hours of flight instruction in a conventional airplane.

2 The manufacturer sells through a dealer. The dealer helps assemble the ultralight and in a brief course, instructs the pilot in the rudiments of flying, FARs and meteorology.

3 The manufacturer has the dealer require the purchaser of his product to take a comprehensive program of ground and flight instruction that the manufacturer prescribes.

Finger Lakes Airsports in Canandaigua, New York is a business devoted almost entirely to ultralights. It has two grass strips in the midst of upstate farmland, providing a perfect setting for flight training. Its owners, John Farnam and Paul Yarnell, are aware of the responsibility involved in sending someone up in a flying machine. Their training consists of ground schools covering physics of flight, micrometerology, Federal Aviation Regulations, flying skills, and basic aircraft maintenance. Students then receive two to four hours of dual instruction in a *Cessna 150*. Although having only about half the wing loading of the *Cessna*, ultralights and *Cessnas* have similar flight characteristics.

Students then move on to transition training. This includes a review of field assembly techniques, engine operation, and a thorough preflight indoctrination. Next comes taxiing over the strips for several hours to get the student acclimated to the ground handling and pre-takeoff speeds.

The first flights are just short hops off the grass runway. The pilot can make a half dozen touch-and-goes on the 2,000-foot strip. When the student and instructor are satisfied the student's progress is at the proper level, and his skills are adequate, the student then takes a short circuit of the field, and from that point on it's just a matter of practice.

Pilots can rent an ultralight for $25.00 per hour for recreational flying from the Airsports owners. Renting ultralights is a practice that is rare at this time, but is expected to expand rapidly, as many pilots would prefer to rent an ultralight for several hours of flying than to take on the responsibility of owning one.

Airsports charges between $300 and $400 for teaching a novice to fly.

Training

The manufacturer-dealer training programs fall into several general patterns, with some similarities among them. Almost all give ground training—some have established formal classrooms for this purpose. Courses cover technical aspects of ultralights, aerodynamics, theory of flight, micrometeorology and federal regulations. The schools use training films to amplify and intensify the training experience. Students then are given flight in a "simulator," which is a nonflying cockpit device into which the student climbs, takes over controls, and simulates flying an ultralight.

A system used after ground school or in conjunction with it is to have the student begin flight training in his own ultralight or a similar one provided by the school. The instructor works closely with the student, briefing him on what maneuvers to practice, and then running alongside or driving alongside in a car using signals, to assist the pilot to gain skills. Some schools now use compact radios, the instructor talking the student through maneuvers. The student's radio is attached to his helmet.

A tow system of flight instruction has been developed by American Aerolights, manufacturer of the *Eagle*. The dealer used an *Eagle* from which the engine and wings have been removed. A car or truck tows the craft by a 100-foot nylon rope. Once the car starts moving and the rope tightens and starts the ultralight, it is in tow from takeoff to landing. Before the tow starts, the instructor briefs the pilot on limited maneuvers he is to do. In succeeding tows, the novice is given more difficult maneuvers consistent with his progress. Another system in which an *Eagle* is flown in a restricting frame on a platform pushed by an automobile is also used by *Eagle* dealers (see photo, opposite).

Several manufacturers require their dealers to insist that the student take several hours of dual instruction in a *Cessna 150* or similar airplane at a certified flight school before starting to taxi or fly other ultralights. Since two-seat ultralights are not authorized for dual instruction, they consider that this is the only way to get the student to have an opportunity to learn how to use controls while in flight and get the sensation of the three-axes world through takeoffs and landings, stalls, engine stoppages, and emergency procedures. These are all flown before the ultralight instructor allows the student to fly solo in an ultralight. The authors feel this inclusion in a flight training regimen makes sense. In any event, a $4,000 investment in an ultralight, and one's own safety, warrant investment in time and a small additional amount for all training a pilot can get to assure the ability to fly as safely as possible.

Steve Grossruck, president of Cascade Ultralights, manufacturer of the *Kasperwing,* applies a strict flight experience policy: "We manufacture a top-notch, high performance ultralight, and we feel it is our responsibility to insure our customers the highest degree of safety when entering the ultralight field.... It is our

Automobile pushes the *Eagle* simulator up to speeds of 40 miles per hour. The Eagle raises above the platform as the car gains speed. Restraining brackets prevent the craft from raising above 2½ feet from the platform. *Photo courtesy of* American Aerolights

policy to not deliver an ultralight to an inexperienced individual solely because he or she can afford to purchase one." Instead, he refers a potential purchaser to one of his qualified dealers for flying training, or a standard FAA approved flight school for conventional aircraft, or other qualified flight schools for training. He will not deliver a *Kasperwing* to any purchaser unless the purchaser has satisfied Cascade's flight experience requirements. He adds, "The color of the dollar does not a pilot make!"

Learning to Fly Ultralights

The question is often asked: "What if I have experience in some other flying activity?" The answer to that varies, depending upon the amount of experience and the type of airplane. The pilot with a private license and advanced flying experience most likely already knows the regulations, traffic patterns, and meteorology. The hang-glider flyer knows much about weight shift, thermals, and winds. However, he may feel uncomfortable with engine and flight controls. Ultralight flying requires a unique application of hang-glider flying, an engine, surface controls, and much more combined into one craft.

On the other hand, the experienced pilot often has trouble getting used to body movement to shift the center of gravity, plus working the hand controls at the same time. Sometimes this is in opposition to habits he has already learned. Sometimes uninitiated, inexperienced newcomers to flight do equally well as those with a pilot's license because they have no habit patterns or preconceived notions.

Ground Training

Training prior to actually climbing into the cockpit is very important. An ultralight pilot, as well as the conventional pilot, *must* prepare for flight. This includes preflight study of at least the following:

Federal Aviation Regulations (FARs). The air above us is not without its regulations. Part 91 of the FARs contains most of the information the pilot should have.

FARs are important, vital, and should be learned, understood, and complied with.

Aerodynamics. Some ultralights do not handle in the same way as conventional aircraft. Weight shift and unconventional flight control surfaces make the flying different. It is important to understand how an ultralight flies, in addition to learning basic aerodynamics (see Chapter 6).

Meteorology. Weather is important to any pilot. General weather and micrometeorology are the areas to learn. They are explained in Chapter 9. Micrometeorology, the knowledge of air near the ground, is doubly important for the ultralight pilot because much flying takes place at low altitudes, and the ultralight craft is affected much more by light winds than a conventional aircraft.

Local Environment/Pilot's Flying Area. This is an ongoing step in the learning process. Before flying over any new location, the pilot must find out several of its characteristics:

1. What is the terrain, airport, or field like that he is using for takeoff and landing? Are there lakes, trees, open fields, towns, cities? The pilot should determine the direction of prevailing winds and emergency landing areas, among other geographical concerns, and look for obstructions.
2. What other airports, terminal control areas, or prohibited areas are in the flying area? What are the traffic patterns of other local airports and controlled or uncontrolled airports? What is the low-altitude air structure? The answers to these questions are usually easy to get from the operator of the local airport, and it is imperative that a pilot know. Since the ultralight is not the only craft occupying the air above and airspace is regulated, the air must be shared with other craft. A local aeronautical map called a "sectional," showing restricted areas, other airports, and terminal areas may

be obtained from most airports or flying schools. It is an invaluable aid.

Last, but not least, if at a local airport, find out what the local traffic pattern is. Conform to it closely. Also, even if only flying out of a farmer's field, the pilot should establish a similar pattern.

Preflight Inspection

A thorough walk-around inspection of the ultralight—or "preflight," as it is called by experienced pilots—is the first step in starting to fly. The condition of the craft is vital, and the preflight routine is a good habit to get into. Once in the air, it is not possible to

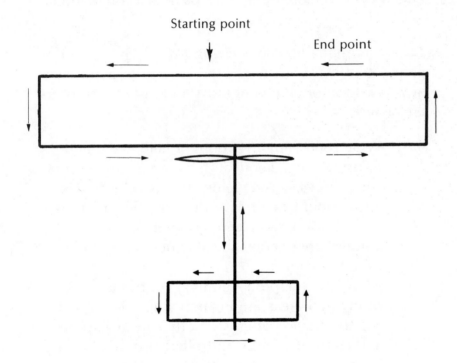

Ultralight preflight check

"pull over to the side of the road" and reconnect a loose wire to its sparkplug, as in a car. In most ultralights, in fact, there is only *one* plug! Therefore, if anything in the craft is loose or amiss, the time to find out is *before* getting into the air.

The walk-around has a starting and ending point. It begins at the nose and proceeds to the right circling the craft, and ending where the inspection started.

Basic things to check are:

loose nuts or bolts	engine/propeller
frayed cables	fuel and lines
wheels	seating attachments
controls	fabric
battens	ignition wiring
pulleys	instruments

Before Starting the Engine

Before attempting to fly, it is important also to check these:

1. Are chocks/tie-down devices removed?
2. Is the helmet securely on?
3. Are straps secure and tight?
4. Is the area clear of people and obstructions?

A habit pattern should be established for these procedures also.

Engine Start

This will vary with the ultralight and the make of the engine. The procedures in the manual should be followed. Here is a generalization of what should transpire:

1. Check ignition off.
2. Recheck fuel lines to see that they are open and clear.
3. Prime the engine.
4. Crack open the throttle.
5. Recheck the area around your craft for people.
6. If other persons are close by, call out "CLEAR PROP!" This is an important step that many leave out. It does two very important things: It lets those in the area know that the pilot is preparing to start the engine and to stay away

from the propeller area; and it alerts bystanders to the fact that the pilot is probably going to be taxiing after the engine starts.
7 Ignition on.
8 Pull starter.
9 After the engine starts, bring it back to idle and let it warm up for two to three minutes. Never attempt flying with a cold engine.

Taxi

"Driving" ("taxiing" is aviation terminology) an ultralight on the ground is not as easy as it might appear to the inexperienced. There are various methods of steering depending on the steering systems in the make of craft. Some have a steerable nosewheel that reacts to foot pressure. Others use air deflection against the rudder as the craft rolls.

Learning to taxi into the wind and with crosswinds is "flying" on the ground and very closely associated with takeoff and landing. It is important to learn this well. Handling a crosswind during taxi can be most difficult because the upwind wing, or that which is facing into the wind, will tend to lift, resulting in the opposite wing tilting downward toward the runway. Depending on how the flight controls are positioned on the ultralight being used, the pilot can counter the crosswind by "flying" the wings, that is, positioning the flight controls to counteract the winds. If there is a crosswind and it is not easy to taxi comfortably, the winds are probably too strong for flying. This difficulty may be sort of a singal to the pilot, a subtle warning that it may be best to call it off for the day.

Takeoff

After a pilot has mastered taxiing, a series of short, low-altitude "hops" is recommended. The short duration is necessary to get experience before going to higher altitudes.

With the ultralight pointed directly into the wind, the takeoff should start with the smooth application of power. When the craft begins rolling, there are several different methods of performing the takeoff, and pilots should consult the flight manual or an instructor. Some use airspeed indicators, and some lift the nosewheel off the ground until the craft begins flying. Keep the wings level with the horizon on either side and accelerate to the best rate of climb speed, the manufacturer's recommended climb speed, or the angle that the instructor recommends. As the pilot gains more experience, this should be a fairly smooth operation, but it will be somewhat uncoordinated the first few times.

Pilots should always prepare for engine failure during takeoff. The techniques for coping with this vary. An instructor will give pointers on what to do.

The pilot should mentally prepare to make an emergency landing during any phase of flight. He should know what the alternatives are. Once off and flying it is hard to convince the pilot to think of emergency procedures when everything is going well.

During takeoff he should ask if there is enough runway ahead to make an emergency landing if the engine fails. What of the field to the left and right? Unless there is sufficient altitude, he will not be able to glide far, and thus will not have much time to make decisions. The idea is this: in the back of the pilot's mind, he must always, in any phase of flight, keep a survey of the terrain below the craft and where he would land if there were a problem. He should remember this as it could save his life!

Cruise

At desired flying altitude, lower the nose until attaining suitable cruise speed, throttle back to cruise revolutions per minute (RPM) noted on the tachometer, and prepare for other maneuvers. It is very important to keep looking *around* the craft, not only straight ahead and down. The point is to "clear" the area for other aircraft. This is a good practice to get into.

Turns

As explained in theory of flight (Chapter 6), turns must counteract centrifugal force and gravity. The pilot will have to increase the angle of attack and advance the throttle to keep airspeed constant. If there is a rudder in the ultralight, and foot pedals or other mechanisms, he will have to push it in the direction of the turn to get a coordinated turn and avoid slipping. A novice pilot should fly a very shallow turn of 10° before increasing the angle. Remember, as mentioned in Chapter 6, stall speed increases with the angle of attack.

Importance of Feel

Almost all conventional airplanes have a nose, instrument panel, and window frames. These are useful for monitoring altitude by eye. Ultralight visual cues are not so apparent, but there are some. If the wings are on the horizon, one can tell if the craft is flying level or not in a turn; and a spot above the nosewheel on the center supporting strut indicates whether the ultralight is climbing or descending. However, this can present a problem to the novice until he learns to replace visual dependency with a physical awareness of changes in the slipstream velocity (airspeed) and the responsiveness of the ultralight controls.

Since these craft have very little weight, they will not retain energy—speed—like a large airplane. Thus, nosing up slightly will immediately result in diminished airspeed. As with larger planes, ultralights become sluggish in the slower portion of their speed range. Mushy, slow responses to controls tell the pilot he is flying slowly in a nose-up attitude. If it feels windier and the controls are crisper, he is in a faster, nose-down attitude.

Most of the aesthetic appeal of ultralight flight comes from the fact that much of it is done by "feel," the sensation of the wind striking the face, rather than by staring at an array of instruments in an enclosed cockpit. The pilot must develop this feel and learn to depend on it. Nevertheless, it is still a good idea to fly with an airspeed indicator.

Most ultralights stall reluctantly in any but the most forceful situation. Stalls are normally preceded by mushiness, sluggish controls, a high sink rate, and a nose-high attitude. Stall-recovery procedures vary somewhat from one craft to another, but normally the pilot will try lowering the nose and increasing power, thereby increasing airspeed. This is a very delicate procedure, so consult the flight manual and instructor. The amount of altitude loss required to get out of a stall varies 50 to 100 feet or more, depending upon the aggressiveness of the pilot's reaction to the stall.

Approach and Landing

The approach and landing are a difficult part of flying an ultralight—or any aircraft. A standard pattern has been established in regulated airports and the pilot is required to fly it. Generally, it takes the form of that shown in the accompanying illustration.

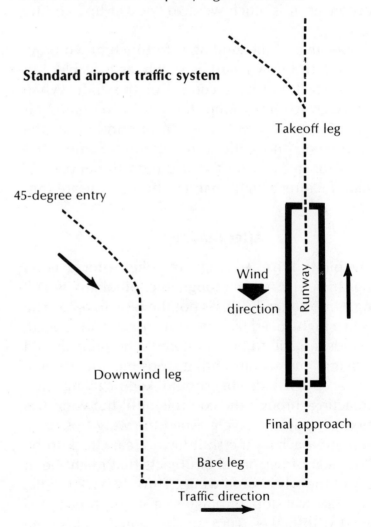

Standard airport traffic system

By using a standard traffic pattern, unless otherwise specified by local airport procedures, a pilot sets up an approach for landing. It is common to overcontrol the craft and feel uncomfortable during first landings. The key is to conduct a stabilized, consistent approach. Try to establish the correct airspeed and glide path as soon as possible on a final approach. To make corrections, adjust the power and angle of attack. If the pilot feels his position in the landing pattern is not right (too fast an airspeed or too high an altitude), he smoothly applies power, raises the nose, and goes around for another approach. It is better to start over again rather than try to salvage a bad approach. Even experienced pilots are sometimes "bullheaded" and try to salvage a bad approach. This is poor judgment under the worst circumstances. It is much wiser to "power up" and try again.

The objective of the landing is to touch down moving forward and downward as slowly as possible, but still be "flying" and have control of the craft. When close enough to the ground, the pilot can "flare" his craft, or level it a few feet above the ground, break the rate of descent, and settle to the ground. Some ultralights do not need to be flared during the approach. The manufacturer's flight manual should indicate this.

After Landing

Many pilots breathe a sigh of relief after a good landing and relax even though the ultralight is still moving. Unfortunately, this is not the time to relax. The craft is in a high-speed taxi and is still near flying speed. Crosswinds are still to be countered. The pilot should keep "flying" the aircraft until it stops moving. If there are strong crosswinds, he should keep "flying" the craft, reacting through the controls until the wind has subsided or handlers have grasped the wing tips.

When approaching the shutdown area, check to be sure it is clear of personnel or objects that might be in the way of the wheels and wings. Throttle back to idle, and the craft will stop, or brake to a stop if the craft is equipped with brakes. Turn off the ignition, close the fuel lines if necessary, unstrap, and get out.

Postflight

Many ultralight pilots may not do it, but postflight inspection is good procedure to observe. Doublecheck that the engine is off and walk around the craft the same way as in preflight. Look for any damage or condition that did not exist on your preflight check. It's good to wipe off any oil or other soil from the fabric before it makes permanent stains.

Flight Maneuvers

The material discussed in this chapter cannot be used as a substitute for the manual and the guidance of an instructor using an approved flight training program. There is much to be learned from an instructor that is not covered here. In order to be fully qualified, the pilot must learn a number of other procedures and must be completely familiar with the differences among various models of ultralights. Some important procedures are: steep turns, downwind turns, clearing turns, stalls, spins, short-field takeoffs and landings, and others. There are books that will enable the pilot to speed his flying education. They are listed in the Bibliography.

Beware of Engine Failure

Two-cylinder engines give greater assurance against total engine failure than one cylinder engines. Although a two-cylinder engine, when a single sparkplug fails, can usually still operate, it is at reduced power, which may only be enough for a controlled descent. Whichever you have, either can suddenly quit. Therefore, adopt the policy of always looking for and keeping in gliding distance of a suitable landing area.

An Instructor Is Best

The authors heartily endorse and highly recommend a manufacturer-approved-and-operated flight-training program. It will be carried out through a dealer, who in some cases will also be the instructor, but *having a professional instructor is a must.* The best programs have a simulator, a craft towed behind a car, or some other similar method. Although it sounds repetitious, the longer the instructor can control a student's flying, the more he or she can teach, and the better prepared to fly solo a student pilot will be.

Pilot Flight Logbook

It is common practice among professional pilots to keep a flight logbook. It records such important data as the date of a flight, the origin and destination, starting and landing times, the airplane, and events of significance. It becomes the pilot's personal flight history. From it he garners knowledge of his total flying time. It shows how current he is in flying and the models of airplanes he can fly.

Although there is no requirement that an ultralight pilot keep such a log, it is wise to do so for the important data it preserves. Moreover, it is possible that the FAA may eventually require ultralight pilots to have licenses, and entries in logbooks may serve, in part, to meet some licensing requirements.

Two other logbooks the ultralight pilot should consider keeping are the aircraft log and the engine log. They enable the pilot to keep tabs on the airframe and the engine and among other things serve as reminders of when it is time to replace parts or major structural members as prescribed by the manufacturer's maintenance manuals.

Insurance

Ultralight flying is such a new sport and statistics are so few that insurance companies have been slow to provide policies. One company, however, Lightwing Insurance, P.O. Box 16, Westerville, Ohio 43081, now offers a package policy that contains liability, physical damage, and medical payments, as well as policies for dealers, ultralight airfields, and professional liability for instructors. The U.S. Ultralight Association, at 1040-A The Almeda, San Jose, California, 95126, also has several policies. USHGA offers liability insurance to its members.

Pat Kubasek, ultralight pilot, stands beside the *Cloudbuster*, a newly introduced ultralight that has conventional, three-axis controls, and a wing design that produces a 15 to 1 glide ratio, one of the highest in the industry. *Photo by the authors*

Pilots are advised to check any life insurance they own to see if it covers ultralight flying, pointing out to the insurance company that no license or certification is required. Many policies do not cover flying during the first two years after the policy is in effect. Some policies cover liabilities caused by accidents. If they do there is no need to get that particular coverage duplicated in any other policy, and unnecessary premiums will be avoided.

Photo courtesy of Boris Popov, Ballistic Recovery Systems, Inc.

8 PARACHUTES

Unhindered, gravity is an irresistible force. It accelerates a free falling object at the rate of 32 feet per second. In other words, at the end of the first second a body is falling 32 feet per second, the next second 32 times 2 or 64 feet per second; the third 32 times 3 or 96 feet per second; consequently, in three seconds it is falling at a destructive 192 feet per second. Thus, gravity becomes a severe problem for the pilot in trouble. When the ultralight engine quits and there is no suitable place to land, or when a flight maneuver puts the pilot in a position from which recovery is impossible, gravity pulls the craft down. A parachute is the logical answer to this situation. Properly used, it helps the pilot to counter the pull of gravity and to recover alive.

As far as parachute use or the wearing of helmets by ultralight flyers is concerned, the sport is in about the same stage as hang gliding was ten years ago. Few hang-glider pilots then wore parachutes or helmets. It was not that parachutes detracted from the "macho" image, but rather that there were no very lightweight parachutes especially designed to meet the special needs of hang-glider pilots. It soon became apparent that many deaths and severe injuries could have been avoided if parachutes had been worn and used in emergencies.

As the sport of hang gliding matured, lightweight parachutes began to appear on the market. As more and more pilots wore them, fatalities started to drop off. In 1978 and 1979, it is estimated that parachutes saved forty lives, halving what would have been the estimated total deaths for those years.

Parachutes clearly increased the overall system safety for the hang glider and pilot, reducing damage to the craft as well as injury to the pilot. However, there was an increase in cost and performance demand on the pilot. The pilot had to be educated in the use of the parachute as well as trained in the mental process of judging when to use it. In addition, he has had to exercise caution that the emergency device did not itself become the problem, which could happen if it were not properly used or maintained. Those who made the extra effort have found that parachutes are a vital piece of equipment.

And so it will be with ultralights. The expected acceleration of popularity will eventually give way to the development of systems that can support the needs of the ultralight, and many lives will be saved.

Unique Circumstances

Parachutes are not commonly worn by ultralight pilots for some of the same reasons that hang-glider pilots did not wear them in early years. But there are other reasons. The use of a parachute in an ultralight presents special problems such as where to install it, how to deploy it, and how to prevent entanglements with structures and engine components. The parachute is deployed through dynamic interaction with the airstream, a portion of which flows through or close to the propeller. And so the propeller presents a more complex design problem for parachutes used in ultralight flying, as contrasted with the hang glider. In addition, ultralights do not generally fly at high altitudes and leave little time to jump if an emergency occurs. Moreover, the pilot in an ultralight is surrounded by a web of tubing and wires, which makes it almost impossible to jump free of the craft if a problem occurs.

As a consequence, and as is the case with all hang-glider parachute drops, the pilot and ultralight will usually drop together as a single unit. This means that an appropriate parachute must be able to sustain the release and free fall descent stresses caused by a 350- to 400-pound load of pilot and craft. Moreover, it is also desirable for the chute to permit the pilot and craft to descend slowly enough so that the landing jolt will not endanger the pilot, and, if possible, will be so slight that the ultralight will not be damaged.

Parachutes in Use

Most parachutes used until recently by ultralight pilots were designed for hang-glider flying, where the drop load (weight of the craft, pilot, and gear) is about 250 pounds. Many chute manufacturers say such chutes will safely withstand drop loads of 400 pounds. While these chutes have proved their worth in hang-gliding flying emergencies, there is no real experience with them in ultralights. To meet the sport's needs, some manufacturers have now developed parachutes that have a 26-foot (or larger) base when deployed, and say these chutes can withstand stresses of 350-pound drop loads at regular ultralight speeds. One manufacturer recently began marketing a parachute for ultralights that has a 573-square-foot area and a quadraconical design. It supports 420 pounds and descends at between 16 to 18 feet per second. Another manufacturer advertises a parachute that is adaptable for use on all current production ultralights with a gross weight less 600 pounds. It usually sells for about $600.

Test the Parachute

Before purchasing a parachute, the pilot should be certain that the dealer will fully explain how the chute works, and stand by to assist if necessary, while the pilot learns to toss it in "Frisbee" fashion from the ultralight, if that is the approved procedure. Any pilot who owns a chute and has not been trained in this procedure should go to the dealer and demand a practice session.

Don't Wait

When a pilot of a recreational, commercial, or military airplane decides to jump because of a malfunction in his craft, he jumps free of the craft and pulls a handle in the parachute harness that causes the chute to deploy. In the ultralight, this is not possible.

Strapped in the harness and slung below the wing, as the pilot of an ultralight is, a much different situation exists when it becomes necessary to use the parachute. The pilot cannot release himself from the harness and drop free from the ultralight, then open the chute, as can the airplane pilot. The pilot and craft drop together when the chute is deployed. Thus, if the ultralight or its appendages (king post and control bar) obstruct the parachute as it opens, the pilot may be multiplying his troubles. Consequently, it is vitally important that the pilot know how to release the parachute pack, grab for one of its four "elephant" ears, and throw it Frisbee fashion as far as he can away from the disabled craft. He must also try to throw it so that the pack does not strike a wire or other obstruction.

Although parachutes, once they are thrown, deploy in a matter of seconds, a pilot and ultralight can fall many feet in these precious seconds before the parachute deploys completely and slows the fall rate perceptibly. The earliest a pilot can deploy the chute, the better.

Sharon Kay Vickery, 26, owner of Vickery's Ultralite Aircraft, specializing in training and sales for *Wizard*, throughout San Bernardino County. *Photo by Don Vickery*

When a pilot has an emergency below 500 feet that seems to require a chute, the consensus among veterans is that quick action to release the parachute is essential. Glenn Brinks says (in *Hang Glider* magazine) that the pilot must take five actions in rapid sequence: look at the chute, tear the Velcro binder open, grab the chute, look for an open patch of sky, and throw it at that patch. With practice, these steps should take two seconds. Chris Price, an experienced hang-glider pilot quips, "When in doubt, whip it out," rather than sacrifice altitude while deliberating whether it is necessary, or attempt to rectify whatever problem has occurred. Since turbulence will sometimes lead the pilot to believe that his craft has developed a problem, Chris would give the pilot no more than three seconds to ascertain whether the craft will rectify itself, and then... whip out the parachute.

What Using a Parachute Can Mean

In mid-1981, Jett Kerby was fatally injured when the right wing of his *Goldwing* folded up and then fell off, allowing the plane to plunge to the ground. Jett was not wearing a parachute. Witnesses estimated the elapsed time between the moment that the wing folded and the plane hit the ground to be only four to five seconds. His altitude when the wing folded was 150 to 200 feet.

Hang-gliding chutes can work from this low an altitude, an experience encountered by Rich Grigsby, a professional hang-glider pilot, who has had to use his chute. A chute just might have saved Jett's life. Even if it had opened only moments before impact, it might have absorbed enough of the momentum to reduce the damage to perhaps a broken leg or hip or some broken ribs.

Military pilots of high-performance aircraft are required to abandon attempts to salvage the craft at 10,000 feet and bail out. A large part of the regulation is based on the premise that the craft can be replaced. The pilot cannot....

The conclusion is obvious: Flying an ultralight without a chute is every bit as hazardous as flying a hang glider without one. The short time available in so many cases makes it very obvious that a chute won't do any good if the pilot doesn't practice getting it out in a big hurry.

Further Development Needed

Development and testing of parachutes to make them more suitable for use with ultralights is continuing. The objective is to produce a bridle that cannot be severed by the propeller if the pilot does not stop the engine before deploying the chute. One manufacturer is experimenting with a tubular nylon strap lined with steel cable, which hopefully cannot be severed by the propeller.

Because much ultralight flying will be done at low altitudes, lower than any aircraft flies, the parachute must be located where the pilot can quickly eject it, and its canopy must deploy quickly.

Evolving Systems

There are several systems now being tested and sold in the marketplace that will significantly reduce the time and ease of deployment. One is produced by Q-Tech of Pretoria-North, South Africa, and is secured to the king post. When the pilot pulls the ripcord handle, the parachute container opens, the pilot chute goes into the slipstream powered by a coiled spring, and then the main deployment bag follows. This system has obvious benefits of being attached to the king post and so is clear of the ultralight's rigging, and also that the pilot does not have to manually deploy the chute.

Another system consists of a military ballistic cartridge with a special projectile attached to the apex of the canopy. When the ballistic charge is set off, the parachute, line, and bridle deploy instantaneously and the chute fills immediately thereafter. This system mounts just above the landing gear of the craft, and the parachute and bridle are discharged downward, moving the chute away from the craft.

A new parachute system has been developed recently in England. The parachute is packed in a two-foot tapered canister which is attached to the keel of a motor-powered hang glider or to a structural member of an ultralight that does not have a keel. In the latter case, the structural member to which the canister is attached must be clear of guy wires and other obstructions that might prevent deployment of the parachute. The pilot can trigger the deployment of the parachute by a line attached to a lever on the canister. The product of a year's work, the chute begins deployment in two-tenths of a second after the pilot jerks the release chord, and when deployed, the parachute lowers both the pilot and the ultralight to the ground.

Care and Maintenance

Manufacturers of chutes provide information on the use, maintenance, and repacking of their products. All strongly advise, "Don't use a parachute pack as a pillow or a seat." Also, keep it away from heat, moisture, and grease. Instructions differ on how often a parachute should be repacked. Some say it should be repacked every month. Others suggest that if moisture gets into the parachute, it should be deployed (spread out manually), dried, and then repacked. As for repacking, be certain that it is done properly. The FAA certifies parachute riggers. A local rigger can usually be found in the Yellow Pages of the telephone book under "Parachutes" or "Parachute Jump Instructions." Dan Poynter's *Parachute Rigging Course* is an excellent sixteen-lesson, home-study workbook for obtaining the FAA senior rigging certificate.

Emphasis should be kept on the ultralight and the skills one learns to fly it. Those skills, properly developed, will reduce the need to depend on a parachute. However, since the unforeseen will always happen when least expected, a parachute is vitally necessary. The parachute has proved helpful in the past and its role in the future of ultralights cannot be ignored.

9 WIND AND WEATHER

Most ultralights should not be flown when the wind is blowing faster than 15 miles per hour. Why not? Because their wing loading is too light, less than two pounds per square foot in most cases. The only aircraft with a lighter wing loading is the hang glider, which averages one pound per square foot. (For an explanation of wing loading, see page 79).

Compare these loadings with that of the *Boeing 747-200F*. Its wing area is 5,500 square feet. With fuel, but without cargo, it weighs 697,000 pounds. This gives it a wing loading of 127 pounds per square foot, which is more than sixty times greater than an ultralight's.

The lighter the wing loading of an aircraft, the easier it is for the wind to affect it, blowing it sideways, slowing it down, or speeding it up if it is a tailwind. The cruising (normal flying) speed of an ultralight is 40 miles per hour. Imagine a wind of 15 miles per hour. If that wind is a headwind (coming at the pilot from the front), the craft will be slowed to a ground speed of 25 miles per hour (40 minus 15).

If the wind is gusty, it will toss a light aircraft around dangerously, demanding the utmost of the control skills of the pilot. Wind gusts to 10 or 20 miles per hour that would not affect a conventional aircraft can toss an ultralight around like a leaf and make it very dangerous for the pilot.

If it is a tailwind, the craft's ground speed will be 55 miles per hour (40 plus 15). The pilot will have to take this wind into consideration on each flight decision he makes, from before he takes off to after he lands.

As an example, an ultralight pilot landing normally has the wind at his back on the first leg of his approach, speeding him up to 55 miles per hour ground speed. On the final leg he is going only 15 miles per hour, since he is heading into the wind. To look at it another way, in a matter of just a few seconds that it takes to make his turn to the final approach he has lost 40 miles an hour of ground speed because of the wind alone!

There are two main factors in determining wind safety values. An inexperienced pilot would do well to fly only during winds of less than 8 miles per hour, so says Jim Zahorik in *Ultralight,* February 8, 1982. Doing so will reduce or eliminate the associated problems of crosswind takeoffs and landings, misjudging wind drift turns, downwind stalls, and wind gradients on approaches to landing. Gust differential is the wind speed difference between the lulls and the peak gusts. If cruising at 25 miles per hour in an ultralight with a stall

Two-place Hummingbird's *Prospector* powered by twin-cylinder (opposed) 25 horsepower Lembach engines. The prospector is one of the highest horsepower ultralights. *Photo courtesy of Gemini International; photo by Ed Sweeney*

speed of 18 miles per hour, and flying into a 15-mile-per-hour wind, if a sudden wind reduction to 7 miles per hour occurs, the airspeed drops to 18 miles per hour, which is stall speed. If flying close to the ground, this is a dangerous development. As Zahorik states, "Wind speed, angle and gusts should be checked for several minutes prior to flight." The authors agree.

Micrometeorology

Wind is one of the elements of weather. Others are the many kinds of rains, clouds, thunderstorms, and hurricanes. The average ultralight pilot is concerned primarily with weather close to the surface of the earth since it will not be often that he ventures to high altitudes. The science of micrometeorology concentrates on weather close to the earth. However, this close earth weather is affected by all other weather.

The Weather Eye

Many ultralight pilots, particularly those who have developed a trained eye for the weather—and their number is large and growing—often are more aware of the qualities of flying weather and can assess the meanings of weather data at low altitudes better than jet or other power-plane pilots, sailplane pilots or balloonists.

The weather at lower altitudes is more relevant to ultralight pilots than to the pilots of other aircraft categories. A pilot of a *Boeing 747* is not seriously concerned with the weather at low altitudes. He can often muddle or plow through marginal to dangerous weather at any altitude, all the while knowing little about its mechanisms because his sturdy craft is built to withstand severe weather forces.

Not so with ultralight pilots. They must know as much as they can about weather. Their craft are feather light in comparison to most airplanes. Moreover, if the pilot aims to turn off his engine to glide or soar, good lift winds must be present. The skilled pilot who wants to get a lot of soaring in for the day must know this from existing broadcasts, weather maps, and satellite photos before takeoff.

The most important elements of weather are wind direction, wind speed, and degree of gustiness. Although wind will also interfere and does mar the overall pleasure, flying can continue between showers. If there is a prediction of rain, especially storms, the pilot might be wise to call flying off for that day.

Obtaining Weather Information

Weather information is available from many sources, most commonly through broadcasts on commercial TV and radio news programs. Now, almost nationwide, noncommercial TV produces aviation-weather forecasts at 6:45 and 8:45 A.M. each morning. The National Oceanic and Atmospheric Administration (NOA) also provides forecasts. That administration's area office will provide the position on the radio dial of the local NOA station. Yet another source is the local telephone company's weather report, which can be obtained with a phone call.

An excellent report prepared especially for pilots is the Federal Aviation Administration (FAA) soaring report. In most areas it is provided by the Pilot Weather Briefing Office. That same office can be telephoned by using an 800 area code, but it is called Flight Service Station. Simply dial 800-555-1212 and request the number for that office from the operator.

The National Weather Service of the Department of Commerce also provides air-weather information. Telephone numbers for this information can be obtained from that department's local Weather Service office. Many pilots now own inexpensive weather radios which enable them to tune in to a number of these weather reports.

To meet the unique needs of pilots, the Aircraft Owners and Pilots Association (AOPA) in Bethesda, Maryland, has set up a computer-operated facility that has a data base of weather and flight information. A pilot can use a special AOPA telephone number and gain access to the system and its data. To communicate with the operator, it is necessary to use a "keypad" similar to a touch-tone telephone's in a way that the computer "understands." AOPA charges a modest fee for access to the computer. A local AOPA office can provide further information.

High and Lows

A basic knowledge of barometric highs and lows is essential for simple forecasting. The shifting of great masses of air, caused by unequal heating of the earth between the equator and the poles, combined with the rotation of the earth, creates the overall pattern of air circulation. In the United States, air generally moves from west to east. In that movement will be two kinds of weather cells, *high pressure* and *low pressure.* High cells alternate with low ones over the surface of the earth.

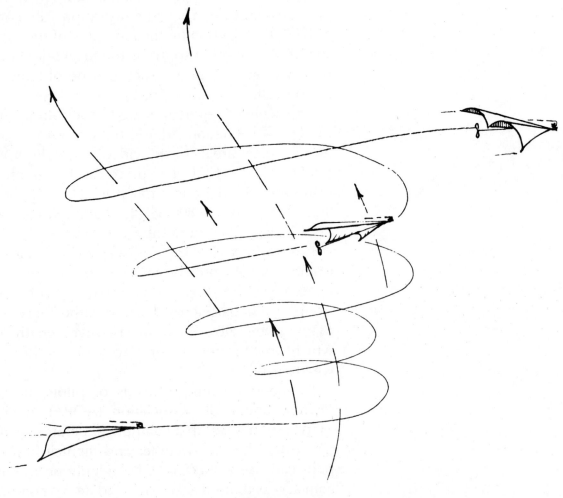

A flexible wing ultralight using thermal lift

Weather conditions

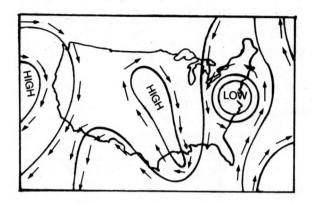
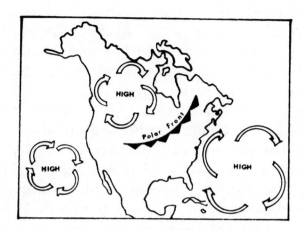

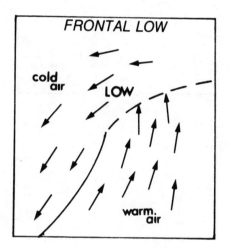

Highs develop where air cools, compresses, and sinks. A high-pressure area is defined as an area of higher pressure surrounded on all sides by lower pressure. Likewise, low-pressure areas are defined as areas surrounded on all sides by higher pressure. As the atmosphere is always seeking a pressure equilibrium, the wind will blow from high pushing to low pressure in an effort to eliminate the pressure difference between the two systems.

As air which has now become wind is pushed out, it is twisted to the right by the earth's rotation. These large, whirling cells repeatedly sweep down from Canada. They are usually several hundred miles in diameter. Their outlines are shown on weather maps as cold fronts.

A *local low* may form when air under a large cumulonimbus (puffy cloud with rain) is rapidly rising. This low pressure area is filled by surrounding air moving in a twisting counter-clockwise direction because of the earth's rotation; it may be twenty-five miles in diameter and can affect surface winds.

A *heat low* develops over deserts and other extremely hot places. The air expands, rises and flows outward higher up. The vacuum formed at the base attracts surrounding air, which rushes in with a swirling motion. "Dust devils" are small versions of a heat low.

Larger ones, lasting most of the summer, form in southwest Arizona and southeastern California. Low-pressure cells can also form on the leeward (downwind) side of mountain ranges and may be large enough to cause local weather disturbances. They often form east of the Rockies in Colorado and Wyoming and in the Texas panhandle. Winds are usually gusty and turbulent in lows.

Highs generally bring fair weather and steady breezes; lows, poor weather and unstable winds.

When you are in the field for several days, you may find it useful to know whether a high or a low is coming. This can be determined with fair accuracy. Stand with your back to the wind, then turn 45 degrees to your right. The high-pressure center is now to your right, the low-pressure center to the left. Since highs always move towards lows, this high will move from your right towards your left. This method is most accurate in the morning, before any local breezes caused by unequal heating have formed. Highs and lows may remain stationary, move quickly, or move in an unusual direction.

These weather signs will provide a basic meteorological guide when you are in the field without access to a weather map.

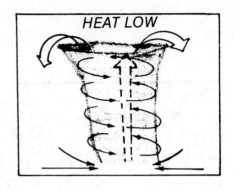

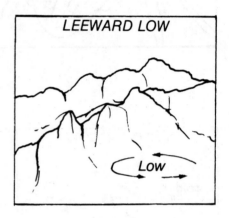

Winds

Wind is one of the most variable of nature's forces. It is not just gusty, strong, or smooth. Rather, it is different at each altitude and in each region of airspace.

While ultralight pilots may experience slope winds, thermals, and other phenomena at high altitudes, most will be flying at slow speeds near the ground, where they will experience *micrometeorology*—small-scale turbulence and air flows created by the terrain. As wind encounters hills and other ground formations and vegetation, it is deflected. Turbulence and wind gradients, which are changes of wind direction and speed, thus develop. These micrometeorological disturbances occur in the surface boundary layer, which extends from the ground to about 170 feet above it.

Minifacts about Micrometeorology

Air behaves much like water, especially as it flows. Water running in a flat sandy bed, whether slowly or rapidly, flows smoothly. Similarly, water flowing slowly in a rock-strewn river may be disturbed little if it moves slowly, but if it flows rapidly, it behaves very differently. It churns as it hits the rocks, flows over and around them and, on their downstream side, froths white with turbulence. As water flows around the rocks, it bends over, or "rotors." Beyond, it eddies and goes into secondary flows.

Turbulence that exists near the ground usually comes from the flow of wind over trees, buildings, hills, and other obstructions. This turbulence becomes increasingly dangerous as the wind moves above 10 miles per hour. Try to visualize wind flow over various obstructions and avoid flight in areas that appear to be hazardous. Stay upwind of possible blocking sources, and if it's necessary to fly leeward of an obstruction, continue as far downwind as possible since rotors and turbulence can continue downwind for miles if the obstruction is large enough, and the wind is high.

Wind Shear

Turbulence resulting from wind shear can occur under many different conditions. Two are especially worth citing, because pilots are likely to encounter them.

One occurs well within the boundary layer. Above a hill or slope pilots frequently encounter a 25-mile-per-hour wind, while the wind at the landing area less than half a mile away is 5 miles per hour or has died down completely. This happens because of the "stickiness" of the layer of air in contact with the ground. Somewhere in between the hills and the landing area, the pilot may experience unexpected turbulence in what should be smooth air. What happens is that a faster-moving body of air passes over the air in contact with the earth, which is not moving. This causes friction which transforms into a form of turbulence.

A pilot may also encounter a vertical wind shear above a hill in the early morning or late afternoon. Sun falling on one side of the hill heats the air, causing it to rise. The other side is in shade, which cools the air, causing it to fall. Shear-created turbulence results, making it extremely dangerous to fly over one of these "shadow lines" in early morning or late afternoon. The best way to avoid doing so is to question fliers with experience about the area and to make map studies to locate areas that might produce vertical wind sheers.

Other Turbulence

So many circumstances cause turbulence that it is not possible to describe them all. The causes of turbulence becomes more complex in the presence of atmospheric stability above the boundary layer. Although high winds occasionally occur and the atmosphere is in constant change, broad segments of the atmosphere, like weather fronts, are relatively stable, even though they move slowly and thus become obstacles to other weather forces. Winds deflected upward by a hill can be turned downward when they encounter a relatively stable weather formation, causing the breaking waves and hilltop-height lee rolls associated with mountain waves. Convection currents caused by warm air moving upward can stimulate local flows, which may "fight" with the mean flow of air in the vicinity and produce wind sheer and turbulence.

Behind a wide, large building, for example, a downward wake can extend in a downwind bubble that may be as high as four to six buildings, with strong turbulence persisting even higher. Doctor Paul MacCready, Jr., states in *Hang Gliding* magazine that since the eddy regions have reverse flows, "the velocity change can be nearly twice the mean wind, as a downwind flow is pushed aside by an upwind one."

He continues: "To generalize, avoid being downwind of anything sharp, or anything that changes, even smoothly, from flat or upslope to a downward slope." Sharp, according to MacCready, applies to precipitous slopes like those at the Torrey Pines cliffs in California.

MacCready warns that an extremely bad condition

exists at the rear of a hill when its face is being hit by strong winds. He flew a sailplane across mountains under low clouds in France in 1956. At mountaintop level, the wind was 38 miles per hour. On the upwind side of a ridge, he would momentarily get lift of almost this much, followed randomly by a downcurrent of equal strength from an eddy behind another mountain a few kilometers upwind. His sailplane was affected by huge G-loads, seriously hampering control and leaving his survival to fate rather than skill.

MacCready concludes that one cannot fly a hang glider safely unless the wind shears and turbulence are considerably less than the craft's slowest flight speed and since an ultralight is almost as lightweight, the caution applies to it as well.

Ultralight pilots, particularly, should look for air that has little turbulence—for example, air coming in from the ocean. Flying in this air should be safer than in air masses traveling the same speed inland. MacCready points out that the topographic features of Torrance Beach, California, if moved two miles inland, would make a far less safe gliding place.

If the upward flow of air is rapid, it will require speed in a landing approach. Flight downwind of sharp obstacles, lee slopes, and hills must be avoided. Where turbulence is strong, flying fast can exaggerate its effects. On the other hand, speed will enable the pilot to get through the turbulence quickly.

To alleviate problems caused by boundary-layer disturbances, more and more ultralight-flying-site managers are setting several wind socks or streamers out over the area, so that pilots can observe the surface winds.

Hang gliding has made major contributions to the knowledge of micrometeorology, but very little is known about wind behavior on airfoils in slow flight at low altitude, that is, between 20 and 100 feet above ground. Ultralight pilots are advised to keep abreast of the latest in a constantly growing body of knowledge about micrometeorology.

The best way for an ultralight pilot to get the feel of and learn about the micrometeorological forces that he will encounter is to get instruction in a light plane and, with the instructor at the controls, fly through them.

Gradually the instructor should turn the controls over to the student who should then fly in similar conditions. Even so, this will not precisely duplicate the feel he will have when at the controls of an ultralight, because he will be flying much more slowly than in a light training airplane.

It is possible that ultralight simulators will some day be able to incorporate devices that simulate wind and turbulence forces similar to those a pilot will encounter, and will have controls that enable the pilot to make corrective responses. But no such simulator has yet been developed.

Until much more is known about micrometeorology and how the pilot can learn by training in simulators to react to the low-altitude weather forces, the old adage, "experience is the best teacher," still applies. This unfortunately must largely govern the manner by which an ultralight pilot senses the feel of turbulence forces and develops proper control responses to them. The authors recommend *Ultralight Airmanship* by Jack Lombie (see Journals and Publications) for additional reading in weather, especially on the subject of turbulence.

10 THE ULTRALIGHT AS A TOW AIRCRAFT

The First Ultralight/Hang Glider Tow

Towing hang gliders with ultralight undoubtedly will rank as a major advancement. This is so primarily for two reasons. First, it has been a leap ahead in flight technology. Second, it opens enormous new possibilities for the sport of hang gliding.

The achievement was bound to occur. Speedboat-, auto-, winch- and other hang-glider-tow methods had established hang-glider towing as feasible and fun to do. Using an ultralight as a hang-glider-tow vehicle merely was to add a new mode of power to get the hang glider in the air under tow.

History provides considerable precedent. Airplanes began towing gliders into the air sixty years ago. As time passed gliders grew larger. Some carried passengers.

Taking cues from the Germans in World War II who had used gliders to land troops in combat, the United States launched a massive assault-glider program. It produced more than 14,000 gliders that carried from fifteen to sixty men each. The gliders carrying glidermen, artillery, and munitions were towed above many enemy strongholds, then released, to land and discharge their men to attack the enemy.

A *Pterodactyl Traveler* tows a *Fledgling* hang glider above the Pacific Ocean. *Photo courtesy of* Glider Rider

Power and Towing

The mating of the hang glider and ultralight waited only for an ultralight that had enough power, structural strength, and integrity to perform the towing task. In July 1981, the *Ascender*, a *Pterodactyl Fledgling* ultralight nicknamed the "Pfug," with Jack McCornack at the controls, took off from an unprepared Everglades airstrip in Florida. Behind trailed a 70-yard nylon rope attached to a *Fledgling* glider piloted by Rex Miller. Rex grasped the control bar and shouldered his glider. As the rope tightened and quickly raised off the ground, Rex began to move forward into a run. Before he had taken more than three steps, the rope tightened, pulled at his glider, and he was in the air, for the first such tow in recorded history.

Ultralight Towing Advantges

Ultralight towing saves a hang-glider pilot a lot of time. It takes ten minutes to get towed to 3,000-feet altitude by an ultralight. This is a lot less time than it takes to drive or trudge up to 3,000-feet altitude on a mountain, if there is one at hand and launch from there. During the period the pilot would be climbing to the mountain launch site, he would have been towed aloft by ultralight, released, flown all the usable lift, landed and have been on his way again many times.

Ultralight-tow surpasses car- or winch-tow. The pilot can get much higher before release, with much more time after release to find a good thermal or other lift source.

Surplus Power Needed

Because they do not have surplus power, most ultralights now on the market are not suitable to tow hang gliders. As a case in point, according to Jack McCornack in the August issue of *Glider Rider*, the *Pterodactyl Ptraveler* climbs at 400 feet per minute which is the average for most ultralights. The *Ascender*, also a *Pterodactyl* but with the Cuyuna 430D, a more powerful engine, climbs at 1,000 feet per minute. Towing needs this higher power. As McCornack explains:

> A conventional tow plane pulls a sailplane, which weighs about a third as much and has about a tenth as much drag as the tow plane. We're towing hang gliders, and they're relatively close to ultralights in weight and drag (with pilot, of course), and it takes gobs of excess thrust to do it.
>
> If you can't climb around a 1,000 grand [feet] a minute solo, don't waste your time towing. You may get off the ground, but you won't have enough margin for rough air and your turnaround time will be prohibitively long. A tow plane you can't use on thermally days isn't much use at all; you're going to need something you can climb out of the sink with now and then.

Future Possibilities

Ultralight towing will greatly expand hang-glider flying. From a single flight strip, an ultralight could keep many hang-glider pilots flying, towing up one after another. Soon to come will be glider snatches. In this operation the ultralight is fitted with a two-inch steel hook suspended on a pole beneath.

A fifteen-foot diameter loop at the free end of the tow rope is suspended between two ten-foot poles placed about ten feet apart. The other end is hitched to the glider. The low-flying ultralight swoops over the clothesline arrangement, hooks the loop of the tow rope, and snatches the glider into the air.

This is a method similar to that used by the United States Air Force to snatch large troop-carrying gliders off the ground during World War II. As engines become more powerful, the time may come when an ultralight will be towing two or more gliders, either side by side or in tandem. Such activities will be largely experimental for some years after they are initiated. If these unique towing systems can be perfected, glider flying activity could zoom.

Hang-glider towing by ultralights is a method of towing that, while if offers high promise, must be considered a tow technique still in its earliest stages.

11 AMPHIBIOUS ULTRALIGHTS

Many ultralight owners who live near lakes and rivers have added pontoons to their craft, thereby creating many new flying opportunities. No longer need these pilots tolerate uneven grass strips or long, hot runways.

For many pilots a lake is closer than a good ultralight airstrip, and floats a more handy accessory than wheels. Estimates are that there are a hundred and fifty lakes for each airport in the United States, and about two thousand lakes for every airport in Canada.

Clearly, the number of runways for ultralight amphibians are countless! Over cool waters, runways stretch for miles in every direction. All the pilot needs to do is head into the wind and take off. Uncluttered camping, fishing and bathing sites unaccessible or hard to get to even with a four-wheel vehicle are within easy reach of an amphibious ultralight.

Float Construction

Plans costing $25 and ready-made floats costing about $1,000 are being marketed. Floats are made from styrofoam, fiberglass, ballistic cloth, or aluminum. One manufacturer, Ultimate Ultralights, carves a float from a large block of styrofoam comparable to that used in white drinking cups. It then laminates a tough polyethylene skin over the styrofoam. A thin sheet of aluminum is also added to the bottom. The floats are 9.6 feet

Ultralight float planes, Cypress Gardens, Florida. *Photo courtesy of* Glider Rider

long, 23 inches at their widest point, and weigh 21.5 pounds each. When used with a *Quicksilver* and a 170-pound pilot, the floats sink into the water only two inches (referred to as a two-inch displacement).

Florida Pontoons now offers fiberglass floats that weigh 27.8 pounds each and are 7 feet long and 17 inches at their widest point. The company will soon be shifting to Kevlar, a tough ballistic DuPont product.

Cautions

There are some cautions about this sort of ultralight flying. Wearing a life jacket is a must. Also, read up on applicable FARs and be certain to follow local ordinances. On some lakes, where powered boats or float planes are not allowed, it can be assumed that ultralights would also be prohibited.

It is important to have the correct propeller, one that has brass or fiberglass leading edges to protect the propeller if a situation occurs that causes its blades to chop the water. Turning at supersonic speed when hitting the resistant water can cause serious damage to edges and tips on an all-wood propeller.

How They Fly

"Ultimate Ultralights are extremely easy to use in calm to mildly active water," reports Dan Johnson in the July/August 1981 issue of *Whole Air*. "The technique for takeoff," he continues:

> .... involves full power, push back (or back stick) to maximum tail down till the craft begins to climb on step (very simply, the forward half of the float, which has the greatest depth, thereby removing most of the float from the water.) Once on step, which is easily ascertained by the forward movement of what I call the "splash point" on the floats, pull forward (stick forward) to gain rotation speed. After this point, use normal take-off technique.
>
> Landings are accomplished pretty much like conventional wheeled efforts except the need for a

full flare is not so essential. Braking in the water is dramatic in its effectiveness.

Taxiing, while different, is also very straightforward. One tip is that to make the tightest turn at low taxi speeds, you can stick both feet in the water on the float side to which you are turning. A last thought is that in stronger winds (over 10 mph), downwind taxi turns require a great deal of rudder blast, if in fact, this works at all.

In rough water (one foot waves or more) or in strong, especially laminar air, the ultrafloats are not comfortable. Most pilots follow a low-wind rule for all ultralight activity and, if so, this weakness will never be a problem.

The Capsized Ultralight

Having so little weight, an amphibian ultralight is in danger of capsizing. Fred McCallum of Ultra Pontoons cited the special danger a pilot encounters when his craft capsized on water. He described one incident where the pilot was merely taxiing,

> .... A crosswind caught the wing and tail and over she went. The frightening thing is it happens so quickly. As the plane goes over, the wing merely knifes through the water and is gone.... It only takes two to three seconds for the plane to turn over. The pilot only has a moment to get a gulp of air, then he has other things on his mind, such as getting out of his harness or seat belt and perhaps getting rid of his helmet.

Accept the fact that sooner or later due to wind, pilot error or whatever, your ultralight is going to be upside down in the water. When this happens, the average tube and fabric structure goes completely under water ... and hangs upside down from the pontoons. Therefore, the pilot must know how to swim, and there must be a boat in the water and ready to go to the pilot's aid.

Pterodactyl float planes in formation. *Photo courtesy of* Glider Rider

Three-axis control *Wizard* float plane showing rudder pedals and other cockpit features. *Photo by authors*

Ultralight fitted with snow skis. Snow ski models are becoming increasingly popular for recreation, trapping and hunting during winter months. Frozen lakes offer excellent take-offs and landings. *Photo courtesy of Eipper Formance; photo by Steven McCarroll*

Because a likelihood of capsizing always exists for a pilot flying an amphibian, he should learn escape procedures for the craft model *he is flying*. There is a problem with life jackets. Since a jacket causes the pilot wearing it to float to the surface, this very safety factor can cause a problem as it tends to float the pilot up under the seat and flying wires. McCallum says, "The use of a life jacket or personal floatation should be avoided." The authors have found the balance of opinion runs counter to McCallum. Some advise, wear a parachute, be sure to have a quick release seatbelt, and also carry a knife strapped to a leg, just in case there is a need to cut away from gear or to slash a wing fabric overhead.

The pilot should pull off his helmet at once! The helmet will fill rapidly, but even so may capture some air and buoy the pilots head up, making it difficult or impossible for the pilot to dive out and away from the aircraft structure.

The *Gossamer Condor* winning the Kremer Prize for sustained man-powered flight. *Photo by Don Monroe*

12 MANPOWERED AND SOLAR POWERED ULTRALIGHTS

Unique among ultralights are the manpowered and solar-powered aircraft, particularly those developed by Dr. Paul MacCready. They are ultralights in every sense of the word except for the fact that they are not foot launchable.

Although there have been efforts in recent years in many nations to produce a manpowered aircraft, and there were some that made short flights, none was truly successful until August 23, 1977, when the manpowered *Gossamer Condor*, largely based on hangglider design and technology, captured the $86,000 Kremer Prize. It was the first such aircraft to fly a figure-eight course around two pylons a half-mile apart. Although more than ten other human-powered aircraft had previously flown, some more than 1,000 yards, none was able to make the critical 180-degree turns necessary to get around the pylons. One or two had tried such turns, but their altitudes were not enough for the long wings to keep from scraping the ground, and as this happened, the gliders crumpled, years of painstaking effort coming to naught.

The Hang Glider's Contribution

The *Condor* owed its success in large part to the hang glider. It used important components of hang-glider design such as the king post and a large wing area made possible because of the supporting king post, and far fewer ribs than any other manpowered craft. Aluminum tubing and other lightweight metals common to hang-glider manufacturing went into its construction, or suggested the use of other lightweight materials, many of which had already been tried by hang-glider pioneers at one time or another.

Most important to the realization of this achievement were the talents of one man, Dr. Paul Mac-Cready, a hang-glider flyer, champion sailplane pilot, hang-glider builder and aeronautical engineer. Extrapolating from his knowledge of hang-glider technology, he created a large aircraft with a 96-foot wingspan, a 70-pound total weight and a slow flying speed (10 miles per hour). Backed by an enthusiastic staff of hang-glider pilots and ingenious plane builders, he set to work producing a canard craft (with a stabilizer surface in front of the aircraft) that he described as having six sticks and seventy-two piano wires connecting everything to everything else.

Suspended below the huge wing—larger than that of the giant *Boeing 747* airplane—was a pod, wrapped in tissue-thin translucent plastic, that housed the pilot and the bicycle-pedaling gear that turned the propeller to power the craft.

The *Condor* was piloted (and powered) first by a professional bicycle racer who had volunteered his services. When it became necessary for him to get back to racing to earn a living, another bicycle racer, the 120-pound, gangling hang-glider pilot Bryan Allen took over, and it is he who made the historic flight. Between the two they made more than four hundred flights. They had many crackups, which in some cases were devastating to the craft, but none harmed either pilot. The *Condor* was almost totally rebuilt an estimated dozen times until finally one awesome craft, in a six-and-a-half-minute flight over a one-mile figure-eight course, its propellers turning lazily a couple of times a second, flew over the ten-foot hurdle marking the finish line and became the prize winner. In a no less dramatic sequel some months later, the *Gossamer Albatross,* successor to the *Condor,* made an unforgettable flight over the English Channel.

To many historians the achievements of the Wright Brothers at Kitty Hawk, North Carolina, on December 17, 1903, almost three quarters of a century earlier when they made the first controlled, sustained flight in a power-driven airplane, had been surpassed, for humans had finally mastered sustained flight using muscle power alone.

First Solar-Powered Flight

Dr. MacCready's next endeavor was to build a true solar-powered airplane, one that did not have to rely on batteries to store energy to enable sustained flight. The result was the *Gossamer Penguin.* He used photovoltaic cells to convert sunlight to electricity, which turned the Penguin's large propeller.

With his thirteen-year-old, eighty-pound-son at the controls, the *Penguin* made fifty test flights. Then, on August 7, 1980, with Janice Brown, a thirty-two-year-old California schoolteacher piloting, it made the world's first sustained flight at Edwards Air Force Base in the Mojave Desert. It flew for 14 minutes, 23 seconds, touching the sand several times, with its longest sustained flight being one and a third miles. Thanks again to Dr. MacCready, aviation had made another giant leap forward.

While the *Penguin* was being tested, Dr. MacCready already had the advanced design *Solar Challenger* on the drawing board. Completed some three months after the feat of the *Penguin,* Janice took the *Challenger* on its first flight. Driving the craft's 11-foot, double-blade propeller were 16,128 tiny photovoltaic cells that converted sunlight directly into electricity without storage batteries, generating the equivalent of 3 horse power. The craft's wingspan is 47 feet and its wing area is 245 square feet. The craft weighs 195 pounds, can reach speeds of 40 miles per hour, and has flown as high as 30,000 feet. It has a surprisingly high glide ratio of 13 to 1. Thus, when clouds block the sunshine, and the *Challenger*'s power drops to zero, the craft becomes a superb glider, sailing the winds of the

Sunshine powered *Gossamer Challenger* being flown by California schoolteacher Janice Brown. *Photo courtesy of DuPont Company, by Randa Bishop*

sky to a break in the clouds to catch the sun and power up again. Dupont's Mylar® polyester film covers its cockpit and wings. Kevlar® fiber, which is light in weight, yet has exceptional strength, is used on the craft's wing spar, tail boom and landing gear.

On July 9, 1981, the *Challenger* took off from Paris, France, for London, England. Flying at times at 12,000 feet, it landed five and one-half hours later in London. Subsequently, the craft began a series of tours to a number of cities in the United States. On the completion of these tours, the Smithsonian Institution in Washington, D.C. will accept the *Challenger* for its permanent museum collection.

The "Forever Airplane"

The National Aeronautic and Space Agency is now working to produce a plane with eternal endurance, a "forever airplane," patterned on the *Albatross*.

Such an aircraft would be a step in the design and

production of robot-controlled, stratospheric craft equipped with radios, cameras, and scientific instruments. As compared to deep-space satellites, such as the synchronous satellites 22,500 miles in space, this stratospheric forever airplane would orbit the earth at an altitude of about 20 miles (100,000 feet), enabling us to get a closer glimpse of the earth than conventional satellites permit.

The 200-foot wingspan, 250-pound (including payload) craft would be pushed by a 2.5 horsepower, solar-energized electric motor.

13 HOW TO BUY AN ULTRALIGHT

"See It Fly Before You Buy!"

As with the purchase of anything, the buyer must beware. This is especially true in the purchase of an ultralight. They are quite different from any other vehicles.

It does not take highly tuned skills for an individual to be able to drive almost any automobile, all automobiles having as they do a steering wheel, automatic- or manual-shift transmission, and a brake pedal. This commonality of equipment does not exist in ultralights. Some have only weight-shift control, others conventional airplane three-axis controls, while still others have a combination of weight-shift and conventional controls. When combining one of these systems with the starting mechanism, steerable nosewheel (as some ultralights have), and the pilot's own varying level of skills, knowledge, and intelligence, some ultralights may be more difficult to fly than others. So it is important for the prospective buyer to consider all these factors in selecting an ultralight and not be swayed by his enthusiasm or the influence of a salesperson as to the comparative ease of flying it.

Be certain to find out as much as possible about the history and reliability of the company whose ultralight takes your fancy. Also, most dealers ask for a down payment of as much as a third of the retail price—this may be as much as $1,500. It is good to know that in the sport of hang gliding many manufacturers went out of business within a year or two after they opened their doors. This was not peculiar to that industry. It happened to the auto industry and the aircraft industry in their early years.

In 1932 there were three hundred auto manufacturers in the United States. Nine years later, after the Great Depression, only about ten remained. In the early 1970s there were dozens of hang-glider manufacturers. Now only ten or twelve remain. Undoubtedly the same phenomenon will occur in the manufacture of ultralights. There are many manufacturers now, and it is almost impossible to show all the many different ultralights they have placed on the market.

There are people building and selling ultralights—and very good ones at that—who will declare bankruptcy. This is a cold hard fact! Quality has less to do with this phenomenon in many cases than does a lack of business ability. It will pay to make certain a company can deliver, and deliver *on time* and that it is well established. It is an unfortunate but frequent occurrence that purchasers are promised six-week delivery, put down their money, build their anticipation to a high as the end of the six-week period comes close, and then find that no ultralight is delivered—and more sadly, not even a warning note that production and delivery has slowed down at the factory. But alas, this happens all too often.

Unfortunately, in a sales effort, many manufacturers understate assembly time. Most buyers are not familiar with an ultralight assembly, and there are legions of owners who, instead of taking the stated "40 hours" or "weekend" to assemble their craft, have taken weeks or months. This does not include additional delays caused by having to send back defective parts, or order parts that were omitted from the initial shipment. Moreover, many assembly manuals are not written clearly and lack adequate instructions and illustrations. Before purchasing an ultralight that must be assembled, whether it is a kit or partially built, it is wise to ask its manufacturer to send its assembly manual for study prior to purchase. Or the dealer should show you one to be certain it is clear, easily understood and will not be the cause of prolonged construction time. Talk to other buyers and ask them how long it really took them to assemble the ultralight.

No Certification Required for Ultralights

The government requires general aviation airplane manufacturers to have each new model certified by the FAA, a requirement as yet not imposed on the ultralight industry. Manufacturers have freedom to design, build, and test an ultralight, change and retest it, and market it in a period of months. This keeps down the cost to consumers and helps speed ultralight delivery to eager users.

But there are dangers to the advantages ultralight manufacturers enjoy. In their hurry to get ultralights into the market, they may be overzealous, move too hastily, and market an inadequately tested craft.

Tina Trefethen, product representative and pilot for Eipper Formance, at the side of a *Quicksilver MX* float ultralight. Tina is a hang-glider flyer and holds a qualified aerobatic pilot certificate. She flies and evaluates Eipper Formance's ultralights. *Photo by the authors*

Thus, a purchaser must be very careful that he does not become, in effect, a test pilot for a well *advertised* but inadequately flown or improperly tested craft that may still have some undesirable flight characteristics or structural tubes that may not be sturdy enough to withstand high stresses.

PUMA

The Powered Ultralight Manufacturers Association (PUMA) was formed in early 1981. One of its purposes is to set standards of manufacture and testing. This is a step in the right direction. Its most important impact will be among its members, though this will be some time in developing. There are other foreign and U.S. manufacturers who do not belong to PUMA, and may or may not observe its guidelines.

Thus it is wise for each buyer to learn as much as possible about ultralights and flying them, to study a number of different ultralights that may meet his needs and fit his pocketbook, and to select one that is really what it is advertised to be, and preferably one that is approved by PUMA.

Picking a Dealer

The way to select a reputable dealer in your area is to first contact the manufacturer of the ultralight you are interested in and ask for the name of a dealer in your area. A manufacturer who sells an ultralight directly to a pilot can do little to service it or teach him how to fly it. In going to the dealer recommended by the manufacturer, the dealer should have a store front operation either at an airport or have access to an area in which he can train. Do not select one who operates out of his basement and the back of his van. Look for a dealer with a ground school, a flight training program, and a service program.

Try to get the name of several of his customers, perhaps through the local ultralight club. Ask them questions about the dealer. Does he keep his promises on the delivery of ultralights and spare parts? What is his training program really like? Who teaches it?

The dealer is going to have the fledgling pilot's life in his hands as he teaches him to fly. The new pilot should approach flying with that point in mind rather than getting the cheapest deal.

Newly Introduced Models

Experience with hang gliders has shown that although a manufacturer has considered his hang glider ready for the public to buy and fly, such is not necessarily the case. The public is ultimately the test pilot of any design.

This is a situation unique not only to the pilot public, but also to the automobile-driving and motorcycle-driving public. It is only necessary to refresh one's memory of reports of recalls of thousands of cars of a single model that have had steering or other defects which caused accidents and led to deaths on the highway. In all these cases, the driving public has been the tester. The situation with ultralights is comparable.

Purchasing the Used Ultralight—Cautions

A used ultralight should sell at a lower cost than a new one. If an advertisement on a used one is interesting and warrants further investigation and serious purchase consideration, there are a few important suggestions to follow.

If you telephone first, ask the right questions: "Why are you selling?" "Have there been any crashes or ground loops or props replaced?" Ask about the condition of the engine, covering, tires etc. Ask about everything. Also, make some notes to refer to when you actually see the aircraft.

Look at the craft before you buy it. Look at all the rigging. Check for signs that cables may have been bent sharply or have been twisted. Note whether the rigging has been replaced and, if so, ask why?

Check all the pulleys. Have they been oiled? Do they move freely? Or has a cable been worn flat from just rubbing against a frozen pulley? Look at all the rigging very carefully. It could mean your life!

Note the condition of all the bolts and nuts and other fittings. Has any of the aircraft grade hardware been replaced by hardware store stuff? If you see non-aircraft-grade screws, bolts, nuts, or pins, ask about them. Check the sailcloth thoroughly. Is there noticeable wear? If so, it is a good idea to get the opinion of a sailmaker as to whether it will need replacement or repair.

There is usually a plastic washer between each tube on most of today's ultralights. Look at them carefully for signs of wear or warpage that may indicate unusual stress. Pull some of the bolts out and check for worn spots. Look at the holes for signs off wear or elongation. Look at the tubing to see if it's wearing thin. Thin spots on the tubing are a tip-off that the aircraft may have flown many hours.

If it has low hours, ask why so few? If an ultralight flies well, it should be flying all the time. It is frequently better to buy a used machine with one hundred or so hours on it than one with only six. Long flying time is nothing to shy away from.

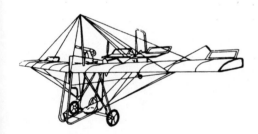

An ultralight that has been flown for awhile has had some engine maintenance. Ask what has been done to keep it flying so long; when the rings were changed; what kind of shape the piston was in? Better yet, look at an old piston. Do a compression test—it should get over 100 pounds even on a cold engine.

Pull a sparkplug. Look for a golden-brown color and for tiny aluminum-colored beads in the end of the plug. Look at the condition of the threads in the cylinder head. Ask if anti-seize compound has been used on the threads.

Look carefully at the wing and tail covers to see if they have been repaired. Is the sail the original? Look at the stitching to see if it all matches. If something looks different ask about it.

Look for nicks, cracks, or deep stress cracks in the propeller. Shallow stress marks in the finish are usually okay for wood props, but thoroughly examine the deep ones. Then, by all means—after getting it home take it all apart. Check everything over very carefully. Replace or repair anything that doesn't look right.

Building Your Own Ultralight

Many manufacturers sell ultralight kits. From kit to completion, in cases where much has been prefabricated, two days of diligent work are all that is needed. For other ultralights, construction can take *five hundred hours or more,* or in terms a bit easier to grasp, 63 eight-hour days, or more than two months of steady labor. This is a lot of time to invest. Many factors can cause this time to extend to a year or more if the builder is not persistent, is not acquainted with handling tools, or does not have the right tools, to mention just a few. It has taken some kit builders as long as two years to complete their ultralight. Enthusiasm runs out quickly. Some models have depressingly low completion rates. It is false economy to think it will save money to build an ultralight from a purchased design. It is much cheaper to get the materials in a kit, rather than to prowl around trying to get a bargain on aluminum tubing.

Thus potential purchasers of kits are advised to check with other buyers of the same ultralight kits they have their eyes on to determine what their actual experience has been, versus what the manufacturer claims.

The ultralight market is highly competitive. Some companies may not survive even though they produce quality ultralights. It is advisable to buy all elements of a kit at one time to forestall disappointment if the company does not survive, in finding that you can no longer get the next section of the kit.

As a builder of an ultralight, an individual is embarking on a major construction project that not only involves time, commitment and money, but also an element of safety. Thus, after examining all aspects of building an ultralight, a potential builder may find it best not to build and better to wait until he can afford a ready-built model.

"Jury-Rig" Repairs

During the first half of 1981, six ultralight pilots were killed in crashes. These deaths resulted from reckless flying and from poorly built craft or craft that were modified incorrectly, according to Doug Hildreth,

Chairman of the USHGA Accident Review Committee.

Ultralight owners are cautioned about making repairs themselves. An investigation report by the manufacturer of a well-known brand of ultralight cited some disturbing facts—facts that all ultralight owners must heed. The make and model of the craft are not important, as the fatal accident could have happened with many other ultralights under similar circumstances.

The crash occurred when a small front-wing fitting (tang) cracked and failed, allowing the wing to fold up at an altitude of about 150 feet. The manufacturer reported that he inspected the plane after the crash and found "several improper repairs, unsafely tied coarse-threated bolts and unauthorized modifications." He added further, "The plane had been crashed seven times by the owner while attempting to fly out of a field which was too small for safe operation. Many parts were bent, re-straightened and re-installed ... the tang which broke wasn't made of factory supplied material either."

"The inspector," according to George R. Chase in Ultralight's "Hotline" (February, 1981), "... felt the probable cause of the accident was poor and improper maintenance procedures, compounded by modified structure which transferred more vibration into the wings. A cluster of tubes in the fuselage structure were bolted to a ⅛" gusset plate which had been added by the builder. This prevented the normal pivoting of the tubes which in turn prevented the normal isolation of vibration."

The manufacturer felt that: "No structural part of an aircraft should ever be straightened and reused. You cannot tell what metallurgical damage may have occurred. Cracks need not be visible for the fatigue life to be adversely affected."

"Foot-Launchability"

If an ultralight design is not foot-launchable, it must be registered as "amateur built," and the person that flies it must have a student's pilot certificate as a minimum. However, the demonstrated foot-launchability of some ultralight models has not been proven, and some manufacturers of these models continue to build and market them. Thus, if a purchaser wants to buy only an ultralight that has been properly demonstrated and the demonstration documented, he should be certain to be shown these documents and insist that he have a copy of each if he buys the model. Then if asked to demonstrate foot-launchability, he merely needs to show the documents as proof and need not go through the foot-launching procedure.

Avoid Pressure to Foot Launch

Although the FAA had a foot-launch requirement and some pilots had the impression they had to demonstrate this on the demand of an inspector, such was not the case, particularly if the ultralight had been demonstrated as foot-launchable by the dealer or manufacturer. Of course, if the ultralight has been modified in any way, that would affect its foot launchability. Dealers or manufacturers can provide photos and other documentations for the pilot to carry so that he could produce them for the inspector and not have to be hassled to demonstrate it himself. It should never be necessary for the pilot to feel compelled to do anything unsafe or unreasonable.

14 MILITARY APPLICATIONS OF ULTRALIGHTS

Ultralights offer unique solutions for many different tasks. Among their many current uses are crop dusting and crop surveys, spotting of schools of fish, and towing hang gliders. Also, their possible use in military operations is being explored by the United States and other nations. Ultralights have already been used in several military operations. Palestinian pilots have flown deep-penetration missions into Israel. Entering the airspace of what is possibly the most heavily defended and electronically sensitive area in the world, one pilot flew undetected 28 miles into Israeli territory, reportedly strafing an industrial area on the way. Israel has bought five *Pterodactyl Ascenders* for training pilots. Saudi Arabia has bought *Quicksilver MXs*. One Middle East nation recently purchased one hundred ultralights.

There are many attractions to ultralights that lend them to military use. They are lightweight, easy to assemble and disassemble, make short takeoffs and landings, can fly from unprepared surfaces, use little fuel, and are difficult for radar to detect. In addition, they are easy to maintain, cost little, and their parts are cheap and interchangeable. They can fly at treetop level or 5,000 feet, and with engines off they can glide in to objectives from high altitudes to make surprise landings. Several ultralight manufacturers have shown interest in redesigning their products to stir military interest.

Redesigns, modifications, and adaptations in other ways for military use could likely consist of camouflaged sailcloth and structural members, stronger structures to withstand an increased wing loading, increased horsepower to accommodate the greater weight coming from added equipment, muted exhaust systems and propellers designed to reduce noise, and a reliable, lightweight communications system. Improvements in mufflers and propellers would enable ultralights to make little noise while flying, thereby reducing detection that would draw enemy fire. At 5,000 feet, ground radar systems will probably not spot the small amount of metal content of the ultralight.

Hit-and-Run Attacks

Battle-ready ultralights should be suitable to carry out hit-and-run attacks. Setting out during darkness, their flight correctly timed, they could climb to within gliding range of their objectives, cut off their engines, and land at dawn without a sound on a road or field adjacent to their objective. Quickly folding down or disassembling their craft and hiding it, if the plan so decided, the pilots could then proceed on foot on their mission, hopefully accomplishing it by surprise. The mission accomplished, they could quickly reassemble each ultralight and in a few minutes be in the air on the way back to friendly lines.

Such operations are feasible since they are quite similar to many carried out in World War II with battle gliders carrying men and guns. There also were similar accomplishments by small aircraft or helicopters since. The use of these craft for low-altitude machine gun strafing of enemy targets, or the dropping of bombs will probably not occur, at least not with the current generation of ultralights. Ultralights are too vulnerable and do not have enough power or load capacity to transport aggressive, heavy weapons or bombs.

Active Combat Ultralights

Modern war is fast paced. Commanders need accurate information about the enemy quickly in an emergency. When possible, this information can be best provided by friendly aircraft. If they are not already overhead to radio information, and this is rarely the case, the front line commander must request "air support" from a higher headquarters. If his radio is dead, or the phone lines are cut, or if there are too many noisy explosions close by, the higher echelon may never get the message. If the message does get through, it will take time, even hours, before a helicopter or light observation plane is over the fighting unit. More time will have to be used to get the pilot oriented and radio him what to look for. So much time may have elapsed that the enemy attack he was to observe and report on will already be over.

Ultralights assigned to the battalion would give the unit commander battleside air support and give answers immediately to almost all urgent questions. A pilot assigned to the unit would make the forward command post his beat. He would have his finger on the progress of the battle, watching the battle from flights overhead and from battle progress maps, and listening via radio communication to the commander and staff. Whenever any tanks were heard or seen, the commander would merely beckon the pilot and tell him, "Take a look!" The pilot would run to his ultralight on a nearby road or field and be in the air in minutes. By the time he was 300 feet up, he would have enough of a view to see the tank attack and quickly radio the number of tanks, their zone of attack, and how deeply they had penetrated. This all would occur in about one fifth the time it would take helicopters or light reconnaissance planes to accomplish the same mission.

As artillery spotters, ultralights should be an artilleryman's dream. Flying above the artillery truck columns in approach marches toward the front lines, the pilot could signal the column's antiaircraft artillery and machine guns of enemy aircraft approaching on a strafing or bombing run. Also, the pilot could radio directions to the lead truck drivers as to roads to take. When the artillery was in position and ready to fire, the ultralight could be alongside the guns, the pilot ready to take off at any time on an observation mission or to "spot" artillery fire or give the location of enemy artillery or infantry targets.

The advantages to the military using ultralights, in addition to those mentioned above, are immeasurable. Ultralights need very little maintenance and only a small supply of spare parts. Any maintenance would be done by the pilot. One truck could carry all the spare parts needed, plus two or three additional folded ultralights. One whose damage would take more than an hour to repair could be pushed into a shell hole and replaced by one from the truck. Cost to the government would be only $4,000. The cost to replace a seriously damaged helicopter is $120,000. Similarly, the military cost to train an ultralight pilot would be very small. It could be done within a division or regiment by an experienced ultralight pilot assigned to the command without sending the student to a flight-training school. This contrasts sharply with the length of time and amount of money it takes to train helicopter and light plane pilots.

A feature of the ultralight that should be attractive to the military is that it uses only slightly more than a gallon of gas-and-oil mixture an hour, something to think about in this era of fuel shortages. Moreover, it uses the regular fuel mixture used by many different military vehicles, which is readily available at any army supply point.

The Air Force could find ultralights useful as a basic trainer. They could be used to acquaint the student with aerodynamics and basic flight and eliminate those who are unsuited to flight, all at relatively little cost.

The Navy might be able to use ultralights as observation craft, artillery spotters, submarine hunters, and basic trainers. Or, imagine seeing a submarine surface and a crew speedily assemble an ultralight on deck, launch it either from the deck or on pontoons, and see it off on a mission. Impossible? Not in the least. Winds permitting, this is an exciting and reasonably feasible possibility.

Fighters

Mitchell Aircraft took a bold step in 1981. Under a cloak of secrecy, it designed and developed a military ultralight, the *XF-10,* which is a much-altered *Mitchell B-10.* It uses the powerful Zenoah G25 B-1 engine, and can streak through the sky at almost 80 miles per hour. Advance materials make it almost radar detection proof. It can carry and deliver several different air-to-air and air-to-ground missiles or bombs.

15 CLUBS, ASSOCIATIONS, AND PUBLICATIONS

Until mid-1981, hang gliding and powered hang gliding (ultralight flying) were considered part of the same sport—hang gliding—for several reasons. Early ultralights were looked upon as hang gliders, since the only difference between the two was that one had an engine and the other did not. The FAA's attitude also strongly influenced this concept in flying an ultralight as a "powered hang glider." Almost all ultralight pilots during the early years of this period had first flown hang gliders and alternated their flying between the two craft; most continue to do so.

But pilots and manufacturers soon began to improve powered hang gliders by adding wheels, controls, and larger engines. These additions, plus better designs, transformed the "powered hang glider" into something like a very lightweight airplane in operation.

Because of its many attractions and advantages, the sport drew many new flyers who had never flown a hang glider and for their own reasons had no intention of doing so.

Clubs Form

In the early 1970s, soon after the sport took hold in America, hang-glider pilots started to form clubs and associations. The United States Hang Glider Association (USHGA) was organized and most of the clubs joined it. Toward the end of the decade, interest and participation in ultralight flying by many members grew. Moreover, the ranks of the Association swelled with new members who flew only ultralights.

Both within and without the Association, ultralight flyers began seeking a stronger voice nationally to represent their vital interests. In 1981 the Experimental Aircraft Association (EAA) formed the EAA Ultralight Association at Hales Corners, Wisconsin, P.O. Box 229, WI 53130. The organization has grown rapidly to where it boasts twenty-four chapters in eighteen states and one chapter in Argentina. The Association published *Ultralight* magazine. One can join the EAA Ultralight Association and receive a membership and its magazine for $15, or for $25, can join the EAA and Ultralight Association together receiving in addition to *Ultralight,* the EAA's publication *Sport Aviation* (see membership card sample).

EAA Ultralight Association membership card

Also in 1981, the USHGA decided to confine its activities to hang gliding only. According to plans, the ultralight segment will split off from the Association on September 1, 1982. It is possible that the ultralight organization that ensues will not affiliate with the EAA Ultralight Association, but with the Aircraft Owners and Pilot's Association (AOPA) in Frederick, Maryland.

Plans are that the new organization will have a rating program, insurance, and perhaps, in time, a national registration and title program.

Bay Area Ultralight Club

This is the story of how one club was formed, and it is typical in some respects of how others are coming into being. Ultralight owners and enthusiasts in the San Jose, California bay area needed an unrestricted place to fly ultralights. Dan Manson and his wife started telephoning and stopping at farms to seek a lease to fly ultralights from some willing farmer's fields. Time after time they were rejected when ultralights were mentioned.

Finally one farmer emerged who didn't think it was a crazy idea, and allowed her land to be used by ultralight flyers. The Bay Area Ultralight Club was formed as a means of collecting the money to lease the land at $150 a month. Twelve flyers paid a $10 initiation fee to get things going and agreed to $10-a-month dues. The average age of the members is thirty-nine. Over thirty craft now use the field.

The club has taken out insurance for the property owner. The gate to the field has a lock to which only members have a key. Any pilot landing his craft on a field of crops is fined $50 by the club.

The club is expanding its activities, and is now planning an education program, including tests for pilots on critical safety problems, among other subjects.

Getting Information

In many parts of America, and in many countries around the world, persons who have casually learned that there is such a sport as flying ultralights may have difficulty getting detailed information about it and where it is active.

Many sources of information exist. A growing number of books about the sport may be purchased, or may be obtained through a library. The periodicals listed in the Appendix are excellent sources of information. Ultralight dealers, who can be very helpful, are usually listed in the Yellow Pages of a city's telephone directory under "Ultralights," or "Flying," or "Flying Services."

Local clubs are more difficult to locate. Few are listed in the directories, not because they are "unlisted" telephone company clients, but simply because so few have permanent offices. Consequently, these clubs' telephone numbers are those of the home or office of the

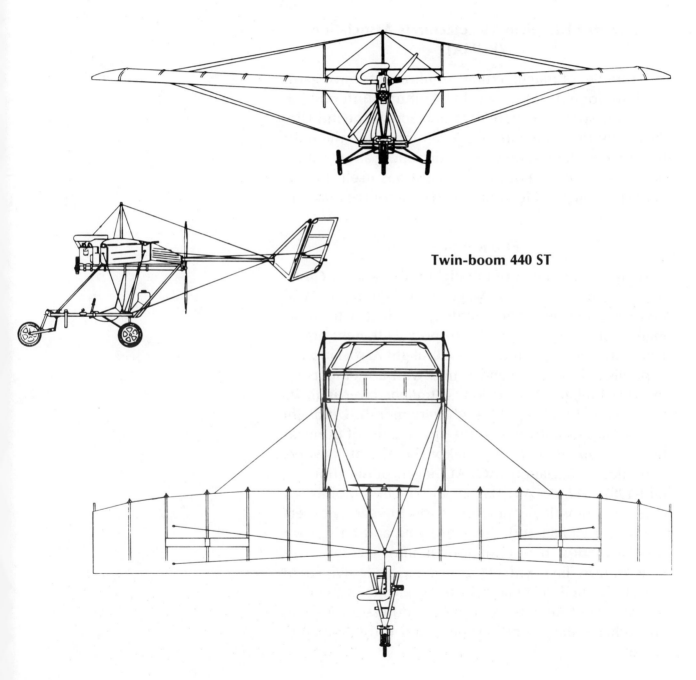

Twin-boom 440 ST

current president of the club. The local hang-glider or ultralight dealer will be the best source for the telephone number and the name of the current president.

Another way to find out about ultralight flying is to telephone the local FAA office, or airport, and ask for the Operations office. Those offices can be very helpful since they are keenly aware of all air activity close by, and if there is any ultralight activity they should know about it.

Powered Ultralight Manufacturers Association (PUMA)

A number of manufacturers of ultralights have gathered to form PUMA, which distributes information about the sport, sets standards, and acts as the industry liaison for the sport with the government and the public. An excellent source of information about ultralights, when it cannot handle questions itself, PUMA generally will provide information on a source that can.

Publications

There are a number of ultralight publications. *Glider Rider,* "The International Magazine for ultralight Aviation," is published every month. It is filled with information on every aspect of the sport. *Ultralight* magazine is the journal of the EAA's Ultralight Association. It is published monthly, and comes with a membership in the Ultralight Association. *Ultralight Flyer* is in its first year and is also published every month. *Ultralight Aircraft* has recently joined the growing list of publications. *Ultralight Pilot,* a journal of the Aircraft Owners and Pilots Association (AOPA), began publications in mid-1982. *Whole Air Magazine,* which is published every two months, is now in its fourth year. It covers both hang gliding and ultralight flying and has grown from a modest newsprint format into a color-cover magazine in line with its growing popularity. *Flight Line,* published in England, is the magazine of the British Microlight Aircraft Association. Addresses for these and other publications can be found in bibliography appendix.

16 ROUNDUP OF ACTIVITIES AND ACHIEVEMENTS

Since ultralight flying is such a new sport, new flying accomplishments and technical advances, along with the organization of clubs and recreational activities, occur almost daily. Perhaps the most phenomenal accomplishments in ultralight technology to date have been the human and solar-powered MacCready *Gossamer* aircraft. They are discussed at length in Chapter 12. While the solar-powered *Challenger* has flown the English Channel and completed other unparalleled feats, there are more over the horizon.

Long-Distance Flights

The powered ultralights, while lesser stars in aviation's orbit, nonetheless have performed well. In 1980, John MacCornack and another pilot flew a Pterodactyl Fledgling across the United States, stopping periodically to refuel and rest. Theirs was a notable flight done at a time when this young sport, ridiculed by other branches of aviation, was struggling for recognition. This one flight vindicated and popularized the ultralight, and proved the durability and air worthiness of this frequently maligned aircraft.

One daring pilot was even on his way to make an attempt at a transatlantic flight with stopovers, until his ardor was dampened by Canadian authorities and his flight finally prevented.

Long-Duration Flights

Records for the time that an ultralight has stayed in the air without landing have been broken regularly. However, a record that will hold for some time was set by Richard J. Britton, an airline pilot flying a *Quicksilver MX* who reportedly flew over Texas carrying extra containers of fuel, from which he periodically refueled the craft in flight. Britton took off early on July 17, 1981. At 3:21 PM, eight hours and twenty-three minutes later, he landed, breaking a three hour record he had previously set.

A more recent U.S. attempt at setting an endurance record was made by Steve Patmont, test pilot for Mitchell Aircraft, on November 4, 1981. Taking off from Porterville's Municipal Airport in California at 8:51 in the morning, he flew nonstop, landing at Porterville at 4:34 in the afternoon, flying a total of seven hours and forty-three minutes (see photo, opposite)! Although he still had fuel remaining, impending darkness forced him to land.

An indication of the progress of the sport are the accomplishments of Gerry Breen, president of Breen Aviation, Inc. In 1979, he flew 200 miles nonstop across Britain. In that same year he flew from London to Paris. In 1980, he flew a trike-powered hang glider 850 miles nonstop from Land's End, England, to John o'Groats at the northern tip of Scotland, and in so doing won the world's first endurance race for ultralights. In 1981 he opened the Microlight Aircraft Division of his company at Enstone Airfield, Oxfordshire, 40 miles from London. Gerry Breen, a dominant name in the sport in the United Kingdom used a trike in his historic endurance race (see Ultralights Around the World, Chapter 17, page 159).

Steve Patmont in his reduction-drive, Zenoah-powered, *Mitchell B 10* just prior to his take off on a seven hour nonstop flight. His arm rests on a 14-gallon fuel tank, one of two carried on the flight. Fueling to only half the 28-gallon capacity, he still had fuel remaining when darkness called an end to his flight. He plans to better the 8-hour and 23-minute record set by Richard J. Britton. *Photo by Howard Waters*

Altitude Records

On February 17, 1981, after coordinating his flight plan with Milwaukee's Air Traffic Control, John Moody took off flying his *Easy Rider* to try for an altitude record. He was aiming to beat his previous 11,100 feet record. At 11,500 feet, with gas to go, he continued climbing. Ninety-eight minutes after takeoff he reached 13,500 feet above sea level, well above his old record, and decided to descend.

A little over three months later, Jim Campbell's *Pterodactyl Ptraveler*, heavily loaded with a well-padded pilot, a radio, parachute, oxygen tank, and instruments, took off at Lakehurst, New Jersey, in an attempt to beat an 18,000-foot record he had set a few days earlier in a practice flight. At 18,000 feet he had his oxygen mask on. It grew cold. His hands lost feeling as he pushed through 20,000 feet. At *21,210* feet, he started his descent.

Ultralight Parks and Airports

Attesting further to the growing ultralight activity is the birth of small airports devoted primarily to serving these vehicles. Some are small, unused airports that an ultralight dealer or manufacturer has purchased.

Ray Woods decided to convert his own airport from a sales, spraying, and maintenance service to the operation of ultralights. His airport is 23 miles east of Brady and 48 miles north of Great Falls, Montana. Weedhopper of Utah has established its own airport in Plain City, Utah. It has a 1,500-foot paved runway on twenty acres and a hangar that can house many ultralights.

The impressive Skyflight Aerodrome is the Phoenix area's first maintenance, training, hangar, and sales facility built for the exclusive use of ultralight aircraft. The center includes a converted commercial building with 33-foot-wide doors, is used as a hangar for as many as twenty ultralights, and has a specially built 1,-000 feet-long landing strip. The strip is about 75 feet wide and has no obstacles nearby to endanger takeoffs and landings.

The first such facility in Ohio is the Liberty Airdrome, 8 miles east of Streetsboro. Robert Essel set up the facil-

Debbie Champaign, former administration pilot and production manager for Vector Aircraft Corp., standing beside an ultralight at the Lakeland Sun 'n Fun in Florida, March 1981. *Photo by the authors*

ity with the assistance of Chuck Slusarczyk of CGS aviation. It has a 2,900-feet north/south runway, inside storage, repair facility, and outside tie downs for ultralights.

The number of facilities devoted entirely to ultralights may increase greatly as numerous commercial, municipal, or other airports rule that ultralight operations are not compatible with normal traffic and force them to seek their own facilities.

Ultralight Activities Growing

Ultralight activities are increasing over the nation. Annual rallies attended by a few to many thousands of people are becoming common in many states. Rallies feature flybys, races, bomb dropping contests, demonstrations and exhibits.

Flight line at the Sun 'n Fun, Lakeland, Florida, March, 1981. One hundred fifty ultralights registered in 1981; 500 in 1982. *Photo courtesy of* Ultralight Flyer

Ultralight participation in the Experimental Aircraft Association's yearly conventions at Lakeland, Florida, and Oshkosh, Wisconsin, is growing. In 1976, a lone ultralight flew at Oshkosh. The next year, three. By 1978, twenty-four appeared. By 1979, the number swelled to sixty-six, and in 1980, ninety-four made the trek to Wisconsin. At the 1981 Lakeland meet there were 108 ultralights; 159 registered at Oshkosh! In one demonstration forty-one buzzed above the approved area, some in tight formations. *In 1982, 500 registered!*

17 ULTRALIGHTS AROUND THE WORLD

Ultralight flight is booming in many countries in the world under many guises and names and conditions. Some countries, such as West Germany, place heavy restrictions on ultralights, and some will not allow them to fly at all.

International Criteria

Many governments have adopted or proposed a 100 kilogram (220.5 pound) weight limit on ultralights. South Africa allows 120 kilograms (264.6 pounds). The Federation Aeronautique Internationale (FAI), headquartered in Paris, France, gives generous weight and wing-area allowances. It defines an ultralight as: "A vehicle not exceeding an empty weight of 150 kilograms (330 pounds), and a wing area in square meters not less than weight (in kilograms) divided by ten, and in no case less than 10 square meters (107 square feet)."

The "... weight divided by ten ..." clause is included to ensure a light wing loading. The idea is that a lightweight craft with a large wing area (light wing loading) will fly slowly. Although the twin engine, 160 pound French *Cri Cri*, and the 296 pound *Quickie* are small and very lightweight, lighter even than some ultralights, they really do not fit the FAI definition of an ultralight because of their small wing areas, and in

The 240-pound *Quickie* cruises at better than 110 miles per hour. Although its weight is close to that of the average ultralight, it cannot be classed as a true ultralight since it does not meet the foot launchability requirement and its wing loading is about four times that of an ultralight. It is a home-built airplane. Its pilot must have a license to fly. *Photo by the authors*

order to produce enough lift they have to fly very fast and have stall speeds in the 50-mile-per-hour range. American ultralights, by comparison, have stall speeds of about 25 miles per hour or less.

According to these criteria, the 155-pound maximum dry weight proposed for ultralights by the FAA is about 70 kilograms (155 ÷ 2.2 pounds per kilogram), considerably below the FAI's limit of 150 kilograms (330 pounds). The FAI's allowable minimum wing area for this weight is 70 ÷ 10, or 7 square meters. Converted this equals almost 75.4 square feet (7 ÷ .836 meters per square yard X 9 square feet per square yard). The wing areas of the lighter weight American ultralights average 150 square feet. This is about one third larger than the 107 square feet minimum allowable under FAI criteria.

Canada

The growing interest in ultralights in Canada compares with that in the United States. Because of the increase in ultralight flying the Canadian Department of Transportation (DOT) issued proposed regulations for ultralights and hang gliders on April 22, 1981. It considers "ultralight" and "microlight" synonymous and defines an ultralight as an aircraft which is powered and heavier than air. It is designed to carry not more than two persons and has a launch mass of 100 kg (220 pounds) or less and a wing loading of 15 kg per square meter (3.1 pounds/feet) or less, calculated using the launch mass plus the occupant mass weight of 80 kg (175 pounds) per person. "Launch mass" means the total mass of a hang glider or ultralight airplane when it is ready for flight, including any equipment, instruments, fuel, but does not include the mass of the occupant(s) or float equipment. The regulations apply to operating (flying) rules, pilot licensing requirements and aircraft certification requirements.

The French *La Mouette,* a combination of a trike and the double-surface, flexible-wing *Azur* hang glider, making a graceful take off in France. *Photo courtesy of* Glider Rider

The DOT has not published the final regulations, but the Microlight Owners and Pilots Association of Canada (MOPAC, P.O. Box 227, Toronto AMF, Ontario L5P 1B1) considers they will generally follow the proposed ones but with some liberalization. Highlights only are given here: A two place ultralight cannot carry a passenger, except when it is being used for dual instruction. There will be no night flying and day flying must be done in accordance with usual flight rules (VFR). An ultralight must not fly closer than five miles from the center of an airport or in controlled airspace. Pilots must wear helmets when flying. For a student's license plus other lesser requirements a pilot will have to make twenty-five takeoffs and landings, and ten hours flying an ultralight, although the final rules may reduce this to only five hours. All this must be done under the supervision and direction of the holder of an Ultralight Airplane Commercial License.

The aircraft must be registered and have registration markings in the wing as do other aircraft. Pilots must take a written examination and also must sign a declaration that they have no physical defect which would cause an inability to pilot an ultralight.

Canadian-built *Lazair* equipped with floats. Notice that the wheeled tricycle landing gear is strapped to the float superstructure. Because of the added weight of the floats, *Lazair* has substituted the more powerful Austrian ROTAC engines for the standard Pioneers. Note also the two propellers in "parallel" with each engine, a stop-gap innovative expedient to match the engine thrust to the propeller load. Later versions will have a suitably sized single prop for each engine. *Photo by the authors*

Canada boasts at least two manufacturers. Adventure Flight Centers, Inc., in Winnipeg, Manitoba, builds the *SkySeeker*. Ultraflight Sales, Ltd., in Port Colborne, Ontario, has been building the popular *Lazair* (see page 000 for a description) since mid-1981 at the rate of ten per week, a rather substantial production.

West Germany

The Federal Republic of Germany (West Germany) is showing great caution, evaluating ultralights carefully to determine how to integrate them into the mainstream of aviation. It has a two-year evaluation program in which twelve pilots are involved. Except for the flying done by these pilots, flying ultralights is against the law, although that is not to say that many pilots are flying them despite the proscription.

British Commonwealth

In the United Kingdom, a lightweight powered aircraft is referred to as a *microlight,* not an ultralight, and its name represents a slightly different concept than its American counterpart. The British choose to use the term microlight to describe what to them is an even lighter weight segment of the more inclusive term ultralight as used for many years in the United States. Interest in microlights is very high. Experimentation and innovation have produced some surprising results, such as the trike (see page 16 for a description), an easily installed power system now being increasingly used for American hang gliders.

The British Microlight Aircraft Association (BMAA) has proposed a definition for microlight, which is patterned on the international definition and gives more generous allowances than the FAA does in its new proposal for American ultralights. The BMAA states the craft is "A single- or two-seat aeroplane having a dry [empty] weight not exceeding 150 kilograms [331 pounds] and a wing area in square meters not less than w/10 [weight divided by 10] and in no case less than 11 square meters [107.6 square feet]" of wing area. The Association considers that these parameters would

Sky Rider designed by Gary Kimberly of Blakehurst, NSW, Australia. It has three-axis aerodynamic controls, flaps, and trim tabs. It is sold in plans only. The engine in the above model is a 12-horsepower, air-cooled, modified McCulloch. *Photo courtesy of Gary Kimberly*

preserve design freedoms while not leading to overfast, structurally weak microlights.

Australia Close to the American ultralight concept is the Australian version, termed a "minimum" aircraft. The Australian Department of Transportation defines such an aircraft as one that has a maximum gross take-off weight of 400 pounds, with a further condition that the wing loading be no more than 4 pounds per square foot. Thus, compared to the average U.S. ultralight, which has an average wing loading of only half that much, the Australian product has the latitude to be much different than its U.S. counterpart.

Minimum aircraft are exempted from normal requirements applicable to commercial aircraft. Pilots need not be licensed, nor must the craft be certified or registered. However, there are many operational limitations, which, as Gary Kimberley states in *Mini-Plane* magazine "means that if you want to fly minimum aircraft, you will have to get out into the country, away from any towns or built-up areas, so that if you have a prang [accident] you're not going to hurt anyone but yourself. The Department of Transport's concern, of course, is public safety."

Minimum aircraft owners and operators formed The Minimum Aircraft Federation of Australia in 1978 and adopted aims and objectives for the sport towards the end that its members fly in a responsible manner so that the government will keep rules and regulations at a minimum.

Except for the United States and England, Australia with a population of only 13,000,000 has more ultralight activity than any other nation. This should come as no surprise. There is a strong background of experience from which the movement has grown. Many world renowed hang-glider pilots are Australian, as is Bill Bennett, owner of the California-based Delta Wing Kites Company. It was Bennett who brought the sport to America, promoted it, and—except for Francis Rogallo, the inventor of the hang glider—gave the sport the thrust that has made hang gliding such a great success in America.

165

Gary Kimberley, who is also well known as a hang-glider pilot has designed and flown the *Sky Rider*, a 195-pound craft that has a maximum takeoff weight of 400 pounds. It has a wing span of 32.33 feet and a wing area of 144 square feet. He has sold more than 130 sets of plans in thirteen countries.

There is a good case for saying that ultralight aviation, as it is now taking form over the world, originated in Australia. This is particularly true in light of Ron Wheeler's introduction in the mid-1970s of the *Skycraft Scout*, discussed in detail in Chapter 3 under Advances in Australia. At one time ahead of the rest of the world with nine manufacturers producing for foreign and domestic markets, Australia continues to be one of the world's leaders.

New Zealand Although New Zealand is well known as a superb country for the flying of hang gliders because of excellent launching sites and an abundance of good winds, ultralight (microlight in New Zealand) flying has been slow in catching on. There are several reasons for this, which also apply in many other countries, unfortunately. One is an unfavorable foreign exchange rate. The New Zealander purchasing a U.S. ultralight must spend nearly $1.20 of his money to buy the equivalent of what an American dollar buys in the United States. Since no ultralights are manufactured in New Zealand, a purchaser must go to sources in the United States or other nations where ultralights are built and sold to purchase one and the unfavorable exchange rate at this time is a big factor in keeping such purchases down.

The second factor is again a cost factor. This is contained in shipping costs and customs duty on ultralights being shipped to New Zealand. Ultimately, an ultralight selling for $4,000 in the United States costs the New Zealander about $6,000 in his currency before it is on hand and ready to fly.

Nonetheless, although few in number, ultralights are now flying. Since the country is largely agricultural, with large sheep farms and ranches supporting some 100,000,000 sheep in the best seasons, and knowing the

Skyercraft Scout, one of the earliest ultralights, is now soaring around the world!

potential uses that farmers there have for ultralights in farm management, their numbers are bound to swell.

The Civil Aviation Administration (CAA) in New Zealand will perhaps take a rather different approach to the regulation of ultralight activity than other nations have. In this respect New Zealand is a leader.

It is the first country in the world to have an approved rating program. It implements this program through a recently formed ultralight association. It is self-regulating, but under CAA guidance. Ultralight pilots must join the association. The association gives flight tests and issues licenses to pilots. It is responsible for ultralight aircraft air worthiness standards. It can actually ground a pilot and in essence, is a self policing organization with freedom from government regulation, the envy of many ultralight pilots in other nations throughout the world.

Japan

Ultralights are becoming popular in Japan. However, Japan with a population of 110,000,000 is no larger in area than California, which has 21,000,000 people. The growth of the sport may level out soon, primarily for lack of space to fly, maintain, and store the craft. Nonetheless, participants, including some U.S. citizens employed there, have organized a club.

18 THE FAA AND ULTRALIGHTS

It is the pilot, and the pilot alone, who is responsible for the safe operation of his aircraft.

FAA Regulation

FAA Definition

On October 4, 1982 the FAA declared effective the long awaited operating requirements for ultralights. Until they were published, ultralight flying was governed by existing regulations for the use of airspace and the 1974 FAA definition of a hang glider described as an "... unpowered, single place vehicle whose launch and landing capability depends on the legs of the occupant and whose ability to remain in flight is generated by natural air currents only." At the time this was published, the FAA urged operators to limit their altitude to 500 feet above ground level; to be alert for aircraft; to avoid controlled airspace and especially airport traffic areas; to avoid flying within 100 feet horizontally of, or at any altitude over, buildings, populated places, or assemblages of persons; and to remain clear of clouds.

Unforeseen Changes Occur

As the sport of hang gliding developed, designers, manufacturers, and operators improved designs and added engines to some hang gliders. Those advances moved well beyond the state of the art contemplated by the FAA. In addition to the use of engines to increase speed, altitude, and distance, operators began to use landing gear and movable control surfaces. Several started to fly two-place or passenger-carrying hang gliders.

The FAA dubbed these new craft "powered hang gliders." However, that agency did not require certification or licensing as long as the craft were foot launched and landed. The FAA did not overregulate the sport. However, it worked with all elements of the sport to promote safety and "self-policing" (regulation).

The freedom the sport enjoyed was beneficial. It stirred creativity in design, manufacture, and operations that enabled the sport to grow phenomenally in a few years. However, by adding engines and controllable aerodynamic surfaces, the designers and manufacturers developed designs closely resembling fixed-wing aircraft, to which published regulations applied.

Irresponsible Acts

As with many other sports, as numbers involved increased, more people began flying ultralights, and "incidents" began to happen. A few unthinking, uninformed, or irresponsible ultralight pilots have flown into regulated airspace, such as airport traffic areas, terminal control areas, control zones, positive control areas, prohibited and restricted areas and federal airways. Some have flown over congested areas and spectators, at night and into bad weather, suitable only for licensed pilots who are qualified to fly under instrument flight rules (IFR).

Typical of what was taking place, on April 11, 1981, a Western Airlines *Boeing 727* captain reported a near-miss with an ultralight vehicle in the vicinity of Phoenix Sky Harbor Airport. On March 24, 1981, an *MU-2* flew between two ultralights operating off the end of the runway at Winter Haven, Florida. Both ultralights were equipped with floats and were operating at night without lights. On yet another occasion, a NASA Alert Bulletin (AB-79-66) described an air carrier flight on downwind for landing at Raleigh–Durham, North Carolina, which flew between two hang gliders without time for evasive action.

Flagrant violations of federal regulations endangered

other aviation traffic in the vicinity, people below, and the ultralight pilots themselves.

FAA Position

Because of the above the FAA was convinced that new rule(s) were necessary. The new rule is 14 CFR (Code of Federal Regulations) Part 103 titled "Ultralight Vehicles, Operating Requirements."* It became effective on October 4, 1982 after prolonged study. It establishes rules governing the operation of ultralights in the United States. The rule defines ultralights in two categories: powered and unpowered (hang gliders).

The intent of the FAA is to provide for safety in the national airspace with a minimum amount of regulation. It considers that the best practices and methods to preclude the need for further Federal regulation appear to at least include: self-regulation and self-policing, safety standards, membership in organizations and associations equipped to function and operate programs approved by the FAA, markings and identification of vehicles, programs including provisions similar to Federal Aviation Regulations relating to aircraft.

The FAA will continue to monitor performance of the ultralight community in terms of safety statistics, growth trends, and maturity and, if indicated, will take additional regulatory actions to preclude degradation of safety to the general public while allowing maximum freedom for ultralight operation. In summary, it should be emphasized that the individual ultralight operator's support and compliance with national self-regulation programs is essential to the FAA's continued policy of allowing industry self regulation in these areas.

Pilot Certification

The FAA endorses the ultralight community's efforts to develop and administer, under FAA guidelines, a national pilot certification program. At this time, however, pilots of ultralight vehicles are not required by Federal regulation to be certified.

---

* *Federal Register* Vol. 47, No. 171, Sept. 2, 1982

Aircraft Registration

Some commenters, primarily state and local governments, recommend that these vehicles be registered and be required to display their registration number. The reasons center around identification of any offenders. The FAA's experience in identification of offenders and processing enforcement action validates their recommendations. The FAA endorses the ultralight community's efforts to develop and maintain, under FAA guidelines, a national registration system which would be immediately accessible to the FAA. However, registration of ultralight vehicles will not be required by Federal regulation at this time.

Aircraft Certification

There are a small number of commenters who recommend additional Federal regulations requiring certification of ultralight vehicles to some design standards. The FAA has consistently refrained from the certification of these vehicles because they were operated by a single occupant for sport or recreational purposes. This policy is in accord with Federal regulatory policies regarding other sport activities. The pilots of these vehicles accept the responsibility for assuring their personal safety much as the driver of a moped street vehicle or a scuba diver does when engaged in his sport. The FAA has noted and commends the efforts of the USHGA to establish design standards and flight testing of new hang glider designs. The FAA endorses the development of similar standards and testing of new powered designs by the ultralight community. However, the FAA presently has no intent to require certification of these vehicles by Federal regulation.

Rules (*Federal Register*, Vol. 47, No 171, Sept. 2, 1982)

General
103.1 Applicability
This part prescribes rules governing the operation of ultralight vehicles in the United States. For the purposes of this part, an ultralight vehicle is a vehicle that:

(a) Is used or intended to be used for manned operation in the air by a single occupant;

(b) Is used or intended to be used for recreation or sport purposes only;

(c) Does not have any U.S. or foreign airworthiness certificate; and

(d) If unpowered, weighs less than 155 pounds; or

(e) If powered:

(1) Weighs less than 254 pounds empty weight, excluding floats and safety devices which are intended for deployment in a potentially catastrophic situation;

(2) Has a fuel capacity not exceeding 5 U.S. gallons;

(3) Is not capable of more than 55 (63 miles per hour) knots calibrated airspeed at full power in level flight; and

(4) Has a power-off stall speed which does not exceed 24 (27.6 miles per hour) knots calibrated airspeed.

103.3 Inspection requirements

(a) Any person operating an ultralight vehicle under this part shall, upon request, allow the Administrator, or his designee, to inspect the vehicle to determine the applicability of this part.

(b) The pilot or operator of an ultralight vehicle must, upon request of the Administrator, furnish satisfactory evidence that the vehicle is subject only to the provisions of this part.

103.5 Waivers

No person may conduct operations that require a deviation from this part except under a written waiver issued by the Administrator.

103.7 Certification and registration

(a) Notwithstanding any other section pertaining to certification of aircraft or their parts or equipment, ultralight vehicles and their component parts and equipment are not required to meet the airworthiness certification standards specified for aircraft or to have certificates of airworthiness.

(b) Notwithstanding any other section pertaining to airman certification, operators of ultralight vehicles are not required to meet any aeronautical knowledge, age, or experience requirements to operate those vehicles or to have airman or medical certificates.

(c) Notwithstanding any other section pertaining to registration and marking of aircraft, ultralight vehicles are not required to be registered or to bear markings of any type.

Subpart B—Operating Rules
103.9 Hazardous operations.
  (a) No person may operate any ultralight vehicle in a manner that creates a hazard to other persons or property.
  (b) No person may allow an object to be dropped from an ultralight vehicle if such action creates a hazard to other persons or property.

103.11 Daylight operations
  (a) No person may operate an ultralight vehicle except between the hours of sunrise and sunset.
  (b) Notwithstanding paragraph (a) of this section, ultralight vehicles may be operated during the twilight periods 30 minutes before official sunrise and 30 minutes after official sunset or, in Alaska, during the period of civil twilight as defined in the Air Almanac, if:
    (1) The vehicle is equipped with an operating anticollision light visible for at least 3 statute miles; and
    (2) All operations are conducted in uncontrolled airspace.

103.13 Operation near aircraft; Right-of-way rules.
  (a) Each person operating an ultralight vehicle shall maintain vigilance so as to see and avoid aircraft and shall yield the right-of-way to all aircraft.
  (b) No person may operate an ultralight vehicle in a manner that creates a collision hazard with respect to any aircraft.
  (c) Powered ultralights shall yield the right-of-way to unpowered ultralights.

103.15 Operations over congested areas

No person may operate an ultralight vehicle over any congested area of a city, town, or settlement, or over any open air assembly of persons.

103.17 Operations in certain airspace.

No person may operate an ultralight vehicle within an airport traffic area, control zone, terminal control area, or positive control area unless that person has prior authorization from the air traffic control facility having jurisdiction over that airspace.

103.19 Operations in prohibited or restricted areas.

No person may operate an ultralight vehicle in prohibited or restricted areas unless that person has permission from the using or controlling agency, as appropriate.

103.21 Visual reference with the surface.

No person may operate an ultralight vehicle except by visual reference with the surface.

103.23 Flight visibility and cloud clearance requirements.

No person may operate an ultralight vehicle when the flight visibility or distance from clouds is less than that in the following table, as appropriate:

Flight altitudes	Minimum flight visibility[1]	Minimum distance from clouds
1,200 feet or less above the surface regardless of MSL altitude:		
(1) Within controlled airspace	3	500 feet below, 1,000 feet above, 2,000 feet horizontal
(2) Outside controlled airspace	1	Clear of clouds
More than 1,200 feet above the surface but less than 10,000 feet MSL:		
(1) Within controlled airspace	3	500 feet below, 1,000 feet above, 2,000 feet horizontal.
(2) Outside controlled airspace	1	500 feet below, 1,000 feet above, 2,000 feet horizontal.
More than 1,200 feet above the surface and at or above 10,000 feet MSL	5	1,000 feet below, 1,000 feet above, 1 statute mile horizontal.

[1] Statute miles. MSL (mean sea level)

Possible Roles for Aircraft Owners and Pilots Association

In an interesting turn of events early in 1982, the Aircraft Owners and Pilots Association (AOPA) formed an ultralight division to integrate this growing segment of flyers into its membership. "Whatever a person flies," AOPA President John L. Baker states, "there is a need for information and understanding of all other aviation activities." His statement was directed not only to ultralight pilots, but also his 260,000 AOPA members, all of whom will gain from the ". . . solving of common problems." AOPA's Air Safety Foundation is working to develop pilot certification standards to ensure safety without undue burden.

The foundation is evolving plans to set up ultralight handling characteristics standards, and is working to ensure minimum structural and construction standards, and devise maintenance procedures. It also plans a basic registration service that will permit safety alerts to be sent, encourage theft protection, and assist recovery of stolen ultralights. The FAA has indicated support for the pilot certification procedures and has further suggested that these procedures may be used instead of federal licensing.

Ignorance of the Law?

A driver is signaled to the side of the road by a policeman for going 35 miles-per-hour in a 25-mile-per-hour zone in a residential community. He has broken the speed limit. He is wrong, even though he did not see a 25-mile-per-hour sign or was totally unaware that a 25-mile-per-hour speed-limit law existed in that community for that road. He can plead ignorance of the law, but ignorance is no excuse, and he will most likely be fined for breaking the law.

Similarly, an ultralight pilot who flies into a restricted area or disregards any other federal or state air regulations is also violating the law. That he did not know the law even existed or that the government requires passing examinations on such laws is no excuse! He should have been responsible enough to have found these laws and not only have *read them,* but *understood* them *thoroughly.*

It is not enough for a pilot to have excellent flying skills and a sound ultralight to fly. He is obligated personally, as well as to others in the air and on the ground, to know flying regulations that pertain to ultralights. The authors want to impress upon every reader of this book that an ultralight flyer must read and understand federal, state, and local flying regulations that apply.

Regulation by States

Some states are taking seriously the need to regulate ultralights. Oregon has a law on its books requiring any powered aircraft to be registered. Moreover, that state is contemplating designating landing and flying areas specifically for ultralights at the state owned airports.

Although ultralights have been used already in some of the above applications or have potential for other applications, the FAA's 14 CFR 103 is clear in that an ultralight "... is used or intended to be used for recreation or sports purposes only." What the impact of this restriction will be on a burgeoning commercial use of ultralights remains to be seen. However, without an early liberalization of the rules, it will certainly be serious, and doom many existing and potential uses.

A squadron of ultralights fly in the evening in Oshkosh, 1982. *Photo by Dan Johnson, courtesy of* Whole Air

19 ABOUT THE FUTURE

The ultralight industry is a beehive of new activity in design and construction. Many new designs are on the drawing board, and some have already gone into prototype and testing stages. Current models are being modified and upgraded continually as well, as experience grows and new technologies and materials appear.

Although thirty models are on the market, new manufacturers are emerging to offer new models every few months. Many other designers and builders across the nation, dissatisfied, with existing models, are producing their own unique ultralights. Among many innovative activities is modification of engines to reduce noise and increase engine lifespan. Weight-shift is giving way to multi-axis controls. Some designers are stressing safety, convinced that the headlong rush by some manufacturers to get designs into production results in the overlooking of important safety factors. From the workshops of these innovators, new ultralights are appearing to move into the lengthening ranks. Some shall ultimately supplant our currently popular models.

Growth is another indication that aviation enthusiasts in general aviation, plagued with higher fuel and maintenance costs and tighter government regulations, are turning to ultralights. Added to this trend is the fact that there is phenomenal growth due primarily to the ultralight's attractiveness as a pure recreational vehicle.

Commercial uses for these versatile small craft are limited only by imagination. Police departments may find them useful for traffic control at trouble spots; the pilot in flight using a bull horn, or radio intercept system to give instructions. The craft may be used in timber cruising and forest disease evaluation, storm damage, and the monitoring of irrigation.

Small farm lots, especially those hard to reach, are well suited for these slow flying maneuverable craft which can take off from an adjoining road or plot unlike many conventional crop-dusting higher speed airplanes which have to fly in from a distant air strip. One experiment showed that a properly equipped ultralight crop duster used only one tenth the amount of chemical spray that a conventional crop duster plane would have used for the same job.

Dan Johnson, publisher of *Whole Air Magazine*, predicts: "Ultralights will go in many directions. Enclosed cockpits for stronger climbs and faster airspeeds. Heavier, smoother, maybe two-seater aircraft. Others will become more simple, packing easier, with lighter weight. They will thermal or ridge soar exceedingly well because that is why someone will build them. And hang gliding will have brought benefits to all aviation...."

One of his predictions has already born fruit. Several ultralights have canvas, plastic or fiberglass enclosures. Other innovations are spoilerons, brakes, anti-stall canards, and ballistically deployed parachutes.

Some ultralights have an engine that can be stopped once the pilot is aloft, allowing the pilot to glide and soar, and then by a pull on a rope he can restart the craft to get to other lifts, or prepare for landing. However, in some designs, this "in-air" restart operation is not easy to handle. This is so primarily because there are still many ultralight engines that are started by hand-turning the propeller. Moreover, although pull ropes start others, these ropes are not accessible to the pilot, once he is seated. For engine-off soaring, many pilots would like a battery and an electric starter. This means an added 6 to 10 pounds for an ultralight.

The Promise of Graphite

Looking ahead, perhaps graphite, a strong, inert substance that is a child of the space age, offers the greatest potential to the glider industry, which is now at somewhat of a technological plateau using aluminum and other currently available materials. Graphite spars and cross tubes, if they are perfected for hang-glider use, would be lighter than aluminum, and the shaving of weight that this would allow could enable higher glider ratios.

Results of some work indicate that graphite tubing is as much as one-and-a-half times stronger and 40 percent lighter than aluminum tubing and about the same stiffness. Ultralights built with graphite parts would be as much as 25 percent lighter than an all aluminum counterpart, a factor that would improve performance.

But the advantages of strength and lightness are only the beginning for graphite. As Eagle Sarmont indicates in the *Whole Air Magazine,* when aluminum first replaced bamboo, the first aluminum gliders to come off the line were bamboo-design gliders merely constructed with aluminum. It was only when designers started using aluminum to its fullest potential that higher performance resulted, as they began to see that aluminum had much more to offer than just stronger structural parts and fewer repairs. As Sarmont says, "It wasn't long before the old rule 'Don't fly higher than you care to fall,' slipped from our minds along with the bamboo bomber designs."

The big disadvantage of graphite is that it costs ten times more than aluminum. Until costs can be brought more closely in line, its use will be limited.

Costs and the Future of the Sport

Unfortunately for the recreational flyer who seeks nothing more than a safe and inexpensive way to fly, ultralights are becoming more expensive, and the reason, in this case, is not inflation. Rather, more powerful, more expensive engines are being used, and designs are better. All of these improvements raise costs, and higher prices in the marketplace result. There is a danger that many who would like to fly may not be able to do so simply because they do not have the kind of money it may take to purchase an ultralight. It is hoped that manufacturers will have at least one machine that they offer that is within reach of the average purchaser.

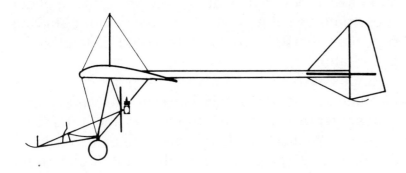

CHAPTERS OF THE EAA ULTRALIGHT ASSOCIATION

ARIZONA
PHOENIX
Robert Cipriano
P.O. Box 32123
Phoenix, AZ   85064
Telephone: (602) 248-8071

CALIFORNIA
BORDERLINE FLYERS
Brad or Cheryl
1452 Stalnacker Rd.
Winterhaven, CA 92283
Telephone: (714) 572-5243

NOR-CAL
Bob Wagner
715 Madison Street
Red Bluff, CA 96080
Telephone: (916) 529-0152

NORTHERN CALIFORNIA
Jim MacVane
P.O. Box 215012
Sacramento, CA 95821
Telephone: (916) 481-5929

PACIFIC COAST EAA #7
Louis L. DuBrul
1116 West Cox Lane
Santa Maria, CA 93454
Telephone: (805) 925-8997

PACOMA EEA #3
Michael S. Michalski
13513 Sayre Street
Sylmar, CA   93424
Telephone: (213) 367-9365

PORTERVILLE EAA #11
Richard T. Avalon
1590 W. Clare
Porterville, CA   93257
Telephone: (209) 781-8411

SACRAMENTO
Joe Manka
UFO Inc.
800 North 10th Street
Sacramento, CA 95814
Telephone: (916) 443-6595

SAN DIEGO EAA #27
Steven M. Reed
7079 Central Avenue
Lemon Grove, CA 92045
Telephone: (714) 460-7040

SAN LUIS OBISPO
R. Michael Rudd
P.O. Box 550
Avila, CA 93424
Telephone: (855) 595-2142

SIERRAS
Emilio Prunetti
P.O. Box 610
Mokelumne Hill, CA 95245

FLORIDA
JACKSONVILLE
Ronald G. Brewer
c/o Rodeway Inn
1057 Broward Rd.
Jacksonville, FL 32218
Telephone: (904) 757-0990

SARASOTA EAA #6
P. D. Clawson
375 W. Baffin Drive
Venice, FL 33595
Telephone: (813) 493-3805

SOUTHEAST ULTRALIGHT
ASSN.
5500 Old Fort Jupiter Road
Jupiter, FL 33458
Telephone: (305) 747-7881
             (305) 747-5129

ILLINOIS
GREATER CHICAGO AREA
EAA #19
Thomas J. Boyle
16408 John Kirkham
Lockport, IL 60441
Telephone: (815) 838-9539

ILLINOIS ULTRALITE AIR
FORCE EAA #34
Frank Beagle
743 S. Osborn
Kankakee, IL 60901
Telephone: (815) 932-9822

ROCKFORD EAA #18
Allen Bednar
625 E. Marshall
Belvedere, IL 61008
Telephone: (815) 547-5126

SACHEM SPRINGS
Eugene P. Forrester II
5344 S. Harper Avenue
Chicago, IL 60615
Telephone: (312) 677-6596

URBANA EAA #30
Dale L. Meadors
70 Gurth
Urbana, IL 61801
Telephone: (217) 367-5020

INDIANA
INDIANAPOLIS EAA #15
Star Brehob
5235 S. Bluff Road
Indianapolis, IN 46217
Telephone: (317) 787-8557

KANSAS
STILWELL
Ken Yadon
7940 Parallel
Kansas City, KA 66112
Telephone: (913) 299-6500

WICHITA
Victoria Barackman
Route 2, Box 93
Wichita, KS 67147
Telephone: (316) 744-2767

LOUISIANA
HOUMA EAA #31
Thomas S. Guidroz
1804 Bambou Drive
Houma, LA 70360
Telephone: (504) 868-1319

MARYLAND
BROOKEVILLE EAA #20
Joseph C. Mitchell
905 Heron Drive
Silver Spring, MD 20901
Telephone: (301) 445-2469

MASSACHUSETTS
ORANGE
Philip Wartel
RFD 251-1
Montague, MA 01351

MICHIGAN
ALMONT EAA #28
Jerry Sandin
3990 Kidder Road
Almont, MI
Telephone: (313) 798-3528

MINNESOTA
ST. PAUL EAA #12
Craig Lamatsch
P.O. Box 7715
St. Paul, MN 55119
Telephone: (612) 441-2858

MISSOURI
ST. LOUIS EAA #16
Chuck Anderson
Ultralight Squadron
850 Hog Hollow Rd.
Chesterfield, MO 63017
Telephone: (314) 469-1033

SPRINGFIELD EAA #21
Jackie A. J. Ott
1703 S. 11th Ave.
Ozark, Mo 65721
Telephone: (417) 485-3362

MONTANA
GLACIER ULTRALIGHT FLYING CLUB
Dennis Granrud
235 Granrud Lane
Kalispell, MT 55901
Telephone: (406) 755-5056

NEBRASKA
LINCOLN
Rollie Woodruff
3168 Puritan Ave.
Lincoln, NE 68502
Telephone: (402) 488-3150

OMAHA
George Yahe
3110 Jo Ann Ave.
Omaha, NE 68123
Telephone: (402) 291-6443

NEW YORK
WESTERN NEW YORK
Gordon Grice
236 Elmwood Avenue
Lockport, NY 14094
Telephone: (716) 434-6731

OHIO
COLUMBUS EAA #25
Doug Smith
2928 Silverton Ct.
Columbus, OH 43227
Telephone: (614) 864-5103

OKLAHOMA
TULSA EAA #10
Terry Boehler
538 S. 101 Ave.
Tulsa, OK 74128
Telephone: (918) 835-1900

OREGON
WILLAMETTE VALLEY
Leonard Tarantola
3920 N. Shasta Loop
Eugene, OR 97405
Telephone: (503) 484-6203

PENNSYLVANIA
DOWNINGTON EAA #14
David Starbuck
752 Norwood Rd.
Downington, PA 19335
Telephone: (215) 269-5109

NEW BRIGHTON EAA #32
Fred Straccia
269 Wises Grove Road, Apt. 3-B
New Brighton, PA 15066
Telephone: (412) 846-9175

SOUTH CAROLINA
HILL'S HOPPERS
Dennis Hill
719 West Smith Street
Timmonsville, SC 29161
Telephone: (803) 346-3151

TEXAS
CORPUS CHRISTIE
Ken Matheson
602 Furman
Corpus Christie, TX 78404
Telephone: (512) 358-0241

FORT WORTH
John E. Davidson
Cowtown Ultralight Association
P.O. Box 4375
Fort Worth, TX 76106
Telephone: (817) 237-1638

WASHINGTON
BELLEVUE EAA #26
Arthur J. Scott
3726 136 Street S.E.
Bellevue, WA 98006
Telephone: (206) 641-6295

YAKIMA EAA #33
Ken Hayward
420 S. 49th Avenue
Yakima, WA 98908
Telephone: (509) 965-3216

WISCONSIN
MILWAUKEE EAA #1
Marvin J. Zellmer
8127 W. Thurston Ave.
Milwaukee, WI 53218
Telephone: (414) 463-1510

WYOMING
U.L.C.
Jerry Alexander
1006 13th Street
Wheatland, WY 82201
Telephone: (307) 322-5557

ARGENTINA
BUENOS AIRES
Gerardo Delamata
659 Segurola
Vincente Lopez, Buenos Aires
Argentina 1639

GLOSSARY

AEROLIGHT *See Ultralight.*

AGL Above ground level.

AILERON Hinged, movable control panel that is set into or near the trailing edge of each wing and usually contoured to the wing. *See Control Surface.*

AIRCRAFT Hang glider, sailplane, or other vehicle that is piloted in the air.

AIRFOIL Wing or other surface shaped to obtain a reaction on itself, such as lift or thrust, from the air through which it moves. Also, the cross-section of a wing.

AIRPLANE Engine-driven, fixed-wing aircraft heavier than air that is supported in flight by the dynamic reaction of the air against its wings.

AIRSPEED Speed, stated in miles per hour or kilometers per hour, of the aircraft through the air.

ALTITUDE Depending on context in which used, the height of the aircraft above the starting point, its height above the ground immediately below, or, as in the case of a commercial airplane, the height above sea level.

ALTITUDE GAIN Maximum altitude achieved above the takeoff point.

ANGLE OF ATTACK Angle at which the air meets the wing surface. *See Pitch.*

APEX ANGLE *See Nose Angle.*

ASPECT RATIO Ratio between the wing span and its chord.

ASI Air Speed Indicator.

AXIS Line through any of the three planes about which an aircraft moves: longitudinal or roll (wing tip to wing tip); lateral or pitch (nose to tail); and vertical or yaw (rudder control).

BANK Tip to the side (incline longitudinally). Attitude when the longitudinal axis is inclined with respect to the horizontal. Roll about the lateral axis.

BATTEN Rod made of fiberglass or similar material that is inserted into a sleeve in the wing so as to stiffen the wing or give it an airfoil shape. Some are preformed into airfoils.

BIPLANE Aircraft with two wings, one above the other.

CABLE RACING *See King Post.*

CANARD A wing protruding in front of the airplane to stabilize it.

CENTER OF GRAVITY Center of weight of an aircraft. If rested on a fulcrum directly under this point, it would balance perfectly.

CHORD Measured from the leading edge to the trailing edge, the width of a wing.

CHT Cylinder Head Temperature.

CONTROL BAR The base of the metal triangle suspended beneath the wing or sail of a hang glider or ultralight. The pilot holds this and exerts force upon it to obtain changes in direction of flight, thus controlling the glider.

COWLING Covering of the engine and sometimes a portion of the fuselage.

CRAB Simultaneous sideways and forward movement through the air.

CONTROL LEVER (S) Attached to aileron rudders, and elevator. Controls action of airplane. (Also *Joystick.*)

CONTROL SURFACE The surfaces that control the rolling, pitching, or yawing. The ailerons, rudder, elevator. Applicable primarily to rigid wing hang gliders.

CONVERGENCE ANGLE *See Nose Angle.*

CROSS BAR *See Cross Tube.*

CROSS TUBE Also Cross Bar. On a Rogallo hang glider, the metal member running through the top of the control bar triangle, which is secured to each landing edge at its extremities. This keeps the leading edges at the proper angle and properly separated.

DIVE Steep forward descent through the air.

DRAG Resistant force exerted in a direction opposite to the direction of flight developed by the wing, sail, pilot, and structural parts of the aircraft.

DRIFT Crabwise motion of an aircraft relative to the ground. Angle between its course and its direction across the ground.

EGT Exhaust-gas temperature.

ENGINE Power source part of power system to promote thrust.

ELEVATOR Hinged, horizontal tail surface which forces the airplane to raise and lower the nose, that is, change the angle of attack. *See Control Surface.*

EMPENNAGE Tail of an airplane including all fixed and movable surfaces.

FAIRING Rigid material shaped to streamline a part so as to reduce drag.

FIN Fixed vertical tail surface of an airplane to which rudder is hinged. Fin gives vertical (yaw) stability.

FOOT LAUNCH "Free-launch" takeoff accomplished solely by the pilot, without the aid of a tow by a boat, automobile, winch, or aircraft.

FUSELAGE The body of the airplane.

G OR G-FORCE The measure of the total force on an aircraft in terms of the force of gravity, for example, 3 G's, meaning that it is under stresses that are three times the force (or pull) of gravity.

GLIDE Literally coasting downhill in the air, gravity furnishing the motive power.

GLIDER Winged, motorless aircraft that flies through the air in gradually descending flight, depending on gravity for power.

GLIDE RATIO Ratio of distance covered horizontally to height lost vertically. *See L/D.*

GROUND EFFECT Cushions of air beneath the wings of an airplane extending to a height of one wingspan above the ground.

HANG GLIDER Defined by the Federal Aviation Administration as "an unpowered, single-place vehicle whose launch and landing capability depends entirely on the legs of the occupant and whose ability to remain in flight is generated by natural air current only." NOTE: Two-place hang gliders are now flying, which otherwise meets the definition.

JOYSTICK *See Control Lever.*

KING POST Aluminum tube protruding above the center of the wing to which cables that support the wing (cable bracing) and the members are secured.

LANDING SPEED Speed at which an airplane touches the ground.

LAUNCH MASS Total weight of an ultralight ready for flight including equipment, instruments, fuel, but not the weight of the pilot, or float or ski equipment.

L/D (LIFT-TO-DRAG RATIO) Ratio of the lift an airplane produces to the amount of drag on it. It determines the glide ratio and is equal to it. *See Glide Ratio.*

LEADING EDGE Front edge of a wing.

LEECH Amount in inches of conical billow in the trailing edge of a sail of a flexible wing ultralight or glider.

LIFT Upward component of force created by a wing, sail, or other aerodynamically designed surface as it moves through the air. Also, the available upward air currents.

LOG Detailed record of each flight made by a pilot or an instructor.

LUFF Fluttering of any part of a sail.

MICROLIGHT Equivalent to ultralight in the United Kingdom.

MINIMAL PLANE Ultralight.

MINIMUM AIRCRAFT Australian equivalent for ultralight.

MONOPLANE Aircraft with a single supporting surface.

MUSHY A condition indicating that the airplane is not responding to the controls. It usually occurs when there is inadequate speed and the plane is in a nose up attitude.

NACELLE Enclosed shelter on an aircraft for the engine. Sometimes referring to the cockpit area. Serves to streamline and protect.

NOSE ANGLE Angle of the nose of the craft in relation to the horizontal plane. Also: Apex Angle, Convergence Angle.

PANCAKE   To drop to the earth from a few feet as a consequence of losing flying speed and/or stalling the craft.

PENETRATE   Airspeed capability which permits an airplane to make forward progress against a given head wind.

PITCH   A turn about a lateral axis so that the nose rises or falls in relation to the tail or rear elements of the airplane.

PORPOISING   Up-and-down motion of an aircraft, which resembles a porpoise swimming through the water. Usually the result of over-control, especially in a glider or sailplane.

POWER LOADING   Weight of the aircraft plus the pilot divided by the horsepower of the engine.

PROPELLER   Wing foil shaped unit used to create thrust. *See Engine.*

REFLEX   Slight upward bend in the rear of the sail. Reflex is the source of pitch stability, operating like the elevator of an airplane canted up slightly.

RIGGER   Specialist who packs parachutes.

ROGALLO WING   Delta- or triangular-shaped flexible wing glider named for Francis M. Rogallo, pioneer of the design.

ROLL   *See Bank.*

ROTOR   Wind that curves downward and begins to rotate. A cross-section of a rotor would look like a watch spring.

RUDDER   Tail panel hinged to vertical fin for directional control in vertical plane. It overcomes yaw.

RUDDER PEDALS   Foot controls to activate the rudder. Rudder on some ultralights are activated by the control stick.

SINK   Area in the atmosphere where there are descending currents of air, which cause the aircraft to lose altitude faster than in still air.

SINK RATE   The vertical downward component of velocity that an aircraft has while descending in still air. Expressed in feet per second.

SOAR   Fly without the power of an engine and without loss of altitude using vertical currents.

SOCK   *See Wind Sock.*

SPAN   Distance from wing tip to wing tip.

SPIN A flight condition, usually unintentional, where the altitude of the aircraft is nose down and the craft is spiraling.

SPOILER A long, narrow plate along the upper surface of a wing that may be raised for reducing lift and increasing drag. It decreases wing efficiency.

SPOILERON A spoiler that combines spoiler and aileron function, generally operated through the foot pedals.

STABILITY Ability or tendency of a glider, sailplane or air plane to return to a normal flying position if hands and feet are removed from or relaxed at the controls.

STABILIZER Fixed, horizontal tail panel that gives stability in the vertical plane and dampens any tendency for the sailplane to yaw.

STALL Loss of smooth airflow over an airfoil resulting in a radical loss of lift causing a mushing descent or abrupt drop of the nose usually caused by excessively high angles of attack.

STICK Flight control device that can activate (depending on the ultralight's design), one or more of the ailerons, elevator, or rudder.

STRUT Bar or rod used as a brace or stress member.

TACTILE FLIGHT Flight depending on the senses as contrasted to instruments.

TAKEOFF The start, or "launch," of a flight.

TANG A metal piece used to connect wires, tubes, or other structural members.

THRUST In ultralights, the forward directed force produced by the power system (including the propeller).

TIP VORTEX Swirling and turbulent air trailing from the wings of an airplane in flight.

TOWLINE Rope, cable, or wire used to pull a glider or sailplane in flight over land or water.

TRAILING EDGE Rear edge of a wing or other airfoil.

TRIM Overall balance of an aircraft while in flight.

ULTRALIGHT Powered, foot launchable aircraft that weighs less than 155 pounds and carries 15 pounds (about 2.5 U.S. gallons) or less fuel and carries only one person.

UPCURRENTS Rising currents of air.

VFR Visual flight rules.
WAKE The turbulent air behind an aircraft.
WASHOUT TUBE Deflects wing tip upward.
WIND SOCK A wind indicator found at most airfields. Cloth cylinder suspended at larger end from a pole to catch wind and indicate wind direction.
WING Supporting surface of an airplane.
WING LOADING The weight of the pilot and empty airplane divided by the total area of the flying surface.
YAW Turn flatly about the vertical axis.

JOURNALS & PUBLICATIONS

AUSTRALIA

Contact
12 Maxwell Rd.
Pagewood, NSW 2035

Mini Plane
255 Woniora Road
Blakehurst, NSW 2221

CANADA

The Kingpost
British Columbia Hang Gliding Association
3523 Wellington Avenue
Vancouver, B.C.
Canada, V5Y 4Y7

Microlight News Bulletin
P.O. Box 227
Toronto AMF
Ontario
Canada, L5P 1B1

ENGLAND

Flight Line
British Microlight
Aircraft Association
20 Church Hill
Ironbridge
Telford/Shropshire
England TFB 7PZ

UNITED STATES

Glider Rider
P.O. Box 6009
3710 Calhoun Avenue
Chattanooga, TN   37401

Ultralight
P.O. Box 229
Hales Corner, WI   53130

Ultralight Aircraft
16525 Sherman Way
Van Nuys, CA   91406

Ultralight Flyer
P.O. Box 98726
Tacoma, WA   98499

Ultralight Pilot
AOPA
7315 Wisconsin Avenue
Bethesda, MD   20015

Whole Air Magazine
P.O. Box 144
Lookout Mountain, TN   37350

Parachute Rigging Course
Para Publishing
P.O. Box 4232-220
Santa Barbara, CA   93103-0232

Ultralight Airmanship
Ultralight Publications
P.O. Box 234
Hummelstown, PA   17036

The following may be purchased from:

Superintendent of Documents
U.S. Government Printing Office
Washington, D.C.   20402

Student Pilot Guide SN 050-007-00476-1 ($2.75)
Flight Training Handbook SN 050-007-00504-1 ($7.50)
Pilot's Handbook of Aeronautical Knowledge
    SN 050-011-00077-7 ($11.00)